基于 Django 的电子商务网站设计

顾翔◎著

清華大学出版社
北京

内 容 简 介

本书是一本介绍如何基于Django框架开发网站的书籍,分4章。第1章是Python、Django发展历史与概要介绍和安装方法,以及HTTP的基础知识;第2章详细介绍了Django基本知识,包括如何启动Django服务、Hello World程序、HttpRequest与HttpResponse对象、setting.py的配置、cookie和session、Django的MTV开发模式框架、Django的模型与数据库的管理、Django的视图管理、Django的模板管理以及基于Python Requests类数据驱动的HTTP接口测试;第3章以电子商务网站为例,介绍电子商务网站的需求、电子商务网站数据Model设计以及用户信息、商品信息、购物车、送货地址、订单、电子支付模块,最后介绍如何建立自定义的错误页面;第4章在第3章的基础上介绍如何构建安全的网站,分别是密码的加密、防止CSRF和XSS的攻击、权限操作的漏洞以及防止SQL注入。

本书可作为准备使用Django框架开发网站、学习接口测试的软件工作人员的学习用书,也可作为在校本科生和研究生的参考用书。

本书封面贴有清华大学出版社防伪标签,无标签者不得销售。
版权所有,侵权必究。侵权举报电话: 010-62782989 13701121933

图书在版编目(CIP)数据

基于Django的电子商务网站设计/顾翔著.—北京:清华大学出版社,2018
ISBN 978-7-302-50512-9

Ⅰ.①基… Ⅱ.①顾… Ⅲ.①软件工具-程序设计 Ⅳ.①TP311.561

中国版本图书馆CIP数据核字(2018)第139861号

责任编辑:白立军 常建丽
封面设计:杨玉兰
责任校对:焦丽丽
责任印制:沈 露

出版发行:清华大学出版社
网   址: http://www.tup.com.cn, http://www.wqbook.com
地   址: 北京清华大学学研大厦A座 邮 编:100084
社 总 机: 010-62770175 邮 购:010-62786544
投稿与读者服务: 010-62776969, c-service@tup.tsinghua.edu.cn
质量反馈: 010-62772015, zhiliang@tup.tsinghua.edu.cn
课件下载: http://www.tup.com.cn, 010-62795954

印 装 者: 北京嘉实印刷有限公司
经   销: 全国新华书店
开   本: 185mm×260mm   印 张: 14.25   字 数:349千字
版   次: 2018年10月第1版   印 次:2018年10月第1次印刷
定   价: 45.00元

产品编号: 078376-01

前　言

在百度百科中对 Python 是这样介绍的：

"Python，是一种面向对象的解释型计算机程序设计语言，由荷兰人 Guido van Rossum 于 1989 年发明，第一个公开发行版发行于 1991 年。

Python 是纯粹的自由软件，源代码和解释器 CPython 遵循 GPL(GNU General Public License)协议。Python 语法简洁清晰，特色之一是强制用空白符(white space)作为语句缩进。Python 具有丰富和强大的库，它常被昵称为胶水语言，能够把用其他语言制作的各种模块(尤其是 C/C++)很轻松地联结在一起。常见的一种应用情形是，使用 Python 快速生成程序的原型(有时甚至是程序的最终界面)，然后对其中有特别要求的部分用更合适的语言改写，例如，3D 游戏中的图形渲染模块，性能要求特别高，就可以用 C/C++ 重写，而后封装为 Python 可以调用的扩展类库。需要注意的是，在使用扩展类库的时候，可能需要考虑平台问题，因为某些扩展类库可能不提供跨平台的实现。

2017 年 7 月 20 日，IEEE 发布编程语言排行榜：Python 高居首位。"

在许多欧美国家，Python 已经成为各大学的基本教学语言。另外，随着大数据与人工智能(AI)技术的兴起，Python 语言在这方面也表现得非常出色。

Python 有以下几个 Web 开发框架，分别是 Flask、Django、Tornado、Bottle、web.py、web2py 及 Quixote。就本人而言，我最喜欢的是 Django 这个框架，虽然掌握这个框架需要学习很多知识，但是 Django 的目的是为了让开发者能够快速地开发一个网站，它提供了很多模块，如 admin。

作者认为刚开始学习 Django 框架，只需要掌握一些最基本的知识，不需要一开始就学习全部知识，只要掌握了这些基本知识，再根据自己的需求学习其他高深的知识，就会变得更加容易了，这正是书写本书的目的：让读者在最短的时间内尽快掌握 Django 框架。建议：如果没有接触过 Django 框架，甚至没有接触过 Web 开发，就可以首选本书。

本书第 1 章是 Python、Django 发展历史与概要的介绍和安装方法以及 HTTP 的基础知识，但是作者在这里没有对 Python 语言进行介绍，如果没有 Python 语言的任何基础，建议通过其他渠道学习掌握 Python 语言后再来学习此书。

第 2 章介绍 Django 基本知识。这里对 Django 的基本知识介绍得比较全面，读者可能看不懂 2.7 节到 2.9 节介绍的知识，只要了解一些概念就可以了，经常使用到的知识会在第 3 章中结合案例进行详细介绍。

第3章结合电子商务网站项目，通过用户信息、商品信息、购物车、送货地址、订单和电子支付这6个模块进行详细介绍。对于其中的每个子模块，通过如何设计url.py、如何开发view.py、模板的设计、接口测试用例的设计以及接口测试代码的书写进行介绍。这里特别需要指出的是，随着软件质量在软件研发中的地位越来越高，并且随着迭代快速响应用户需求的普及，自动化测试显得越来越重要。但是，单元测试代码的繁多以及基于GUI的自动化测试受界面影响很大的原因，这些测试都没能很好地普及，而基于单元测试与基于GUI的自动化测试之间的接口测试在业界却越来越普及。又由于Python提供的Requests类能够更好地配合接口测试的开发，所以作者在本书中对接口测试的技术和实现方法进行了详细描述。作者也是一边书写本书一边书写电子商务网站和接口测试代码，每次程序结构发生变化，都会运行一下以前写好的接口测试代码，以保证新的修改没有影响以前的功能。另外，接口测试的运行速度远远快于基于GUI的自动化测试。本章包含的接口测试用例共49条，全部运行，耗时仅为2.709 499s。

第4章在第3章的基础上介绍如何构建一个安全的网站，分别是密码的加密、防止CSRF和XSS的攻击、权限操作的漏洞以及防止SQL注入。由于第3章对程序进行了很好的封装，所以在这里对产品和测试代码的修改变得更加简单了。正如在本书中作者所写的那样，我们不可能一开始就能书写一个易于维护的好代码，这需要在书写代码的过程中不断地优化。现在敏捷技术提倡研发过程的迭代优化，同样，在书写代码时也需要进行迭代优化。

本书的产品和测试代码放在网站https://github.com/xianggu625/ebussiness上，欢迎下载。

另外，在书写后期，巴特尔、金鑫、刘中秋、任荣哲、万巧、杨军军、叶微及赵院娇对本书文稿进行了校验，在此表示真心的谢意，同时也感谢家人的鼓励与协助，没有你们的支持，本书是不可能在如此短的时间内完成的。

对于本书的内容和产品以及测试代码如果有什么问题，读者可以加我的微信：xianggu0625，另外，我的个人网站是http://www.3testing.com，下面的二维码是我的微信公众号，欢迎各位扫描关注。

顾　翔

2018年2月于上海

目 录

第1章 Python、Django 和 HTTP ………………………………………… 1
1.1 Python 语言 ………………………………………………………… 1
1.1.1 Python 语言概述 ………………………………………… 1
1.1.2 Python 的安装 …………………………………………… 3
1.2 Django 框架 ………………………………………………………… 4
1.2.1 Django 介绍 ……………………………………………… 4
1.2.2 Django 的安装 …………………………………………… 5
1.3 HTTP 概述 …………………………………………………………… 6
1.3.1 HTTP 的工作原理 ………………………………………… 7
1.3.2 HTTP 的请求 ……………………………………………… 8
1.3.3 HTTP 的应答 ……………………………………………… 9
1.3.4 HTTP 的连接性 …………………………………………… 13
1.3.5 HTTP 的无状态 …………………………………………… 15

第2章 Django 基本知识 ……………………………………………… 17
2.1 启动 Django 服务 …………………………………………………… 17
2.2 Hello World 程序 …………………………………………………… 21
2.2.1 直接打印显示内容 ………………………………………… 21
2.2.2 通过文件模板显示内容 …………………………………… 22
2.2.3 文件模板参数 ……………………………………………… 22
2.3 获取参数 …………………………………………………………… 23
2.3.1 通过 GET 方式获取 ……………………………………… 23
2.3.2 通过 POST 方式获取 ……………………………………… 23
2.4 HttpRequest 对象与 HttpResponse 对象 ………………………… 27
2.4.1 HttpRequest 对象 ………………………………………… 27
2.4.2 HttpResponse 对象 ……………………………………… 28
2.5 setting.py 的配置 …………………………………………………… 29
2.5.1 中间件介绍 ………………………………………………… 29
2.5.2 其他配置介绍 ……………………………………………… 30

2.5.3　自定义静态文件 ………………………………………………………… 38
　　　2.5.4　案例 ……………………………………………………………………… 39
2.6　session 和 cookie ………………………………………………………………… 42
　　　2.6.1　session ……………………………………………………………………… 43
　　　2.6.2　cookie ……………………………………………………………………… 45
　　　2.6.3　Django 的用户登录和注册机制 ………………………………………… 47
2.7　Django 的 MTV 开发模式框架 …………………………………………………… 49
2.8　Django 的模型与数据库的管理 …………………………………………………… 50
　　　2.8.1　Django 的数据库 ………………………………………………………… 50
　　　2.8.2　Django 的模型 …………………………………………………………… 51
　　　2.8.3　Django 的后台管理 ……………………………………………………… 55
　　　2.8.4　Django 如何对数据库进行操作 ………………………………………… 57
2.9　Django 的视图管理 ………………………………………………………………… 61
　　　2.9.1　urls.py 中路径的定义 …………………………………………………… 61
　　　2.9.2　方法中显示内容 ………………………………………………………… 63
　　　2.9.3　处理表单 ………………………………………………………………… 63
　　　2.9.4　分页功能 ………………………………………………………………… 65
2.10　Django 的模板管理 ……………………………………………………………… 66
　　　2.10.1　变量的使用 ……………………………………………………………… 66
　　　2.10.2　标签的使用 ……………………………………………………………… 66
　　　2.10.3　过滤器的使用 …………………………………………………………… 72
2.11　基于 Python Requests 类数据驱动的 HTTP 接口测试 ………………………… 75
　　　2.11.1　测试金字塔 ……………………………………………………………… 75
　　　2.11.2　unittest ………………………………………………………………… 76
　　　2.11.3　requests 对象的介绍与使用 …………………………………………… 80
　　　2.11.4　数据驱动的自动化接口测试 …………………………………………… 85
　　　2.11.5　进一步优化 ……………………………………………………………… 89

第 3 章　电子商务网站的实现 …………………………………………………………… 92

3.1　需求描述 …………………………………………………………………………… 92
　　　3.1.1　用户信息模块 …………………………………………………………… 92
　　　3.1.2　商品信息模块 …………………………………………………………… 92
　　　3.1.3　购物车模块 ……………………………………………………………… 92
　　　3.1.4　送货地址模块 …………………………………………………………… 93
　　　3.1.5　订单模块 ………………………………………………………………… 93
　　　3.1.6　订单支付模块 …………………………………………………………… 93
3.2　数据 Model 设计 …………………………………………………………………… 93
3.3　用户信息模块 ……………………………………………………………………… 95
　　　3.3.1　用户注册 ………………………………………………………………… 96

		3.3.2	用户登录	104
		3.3.3	用户信息显示	109
		3.3.4	用户登录密码的修改	130
	3.4	商品信息模块		134
		3.4.1	商品信息的维护	135
		3.4.2	商品概要信息的分页显示	137
		3.4.3	商品信息的模糊查询	144
		3.4.4	商品信息的详情显示	146
	3.5	购物车模块		149
		3.5.1	把商品放入购物车	150
		3.5.2	查看购物车中的商品	155
		3.5.3	修改购物车中的商品数量	161
		3.5.4	删除购物车中的某种商品	164
		3.5.5	删除购物车内所有的商品	165
	3.6	送货地址模块		167
		3.6.1	送货地址的添加与显示	167
		3.6.2	送货地址的修改	175
		3.6.3	送货地址的删除	179
	3.7	订单模块		181
		3.7.1	总订单的生成和显示	182
		3.7.2	查看所有订单	192
		3.7.3	删除订单	197
	3.8	电子支付模块		200
	3.9	建立自定义的错误页面		200

第4章 构建安全的网站 205

4.1	密码的加密		205
4.2	防止 CSRF 攻击		206
	4.2.1	CSRF 攻击介绍	206
	4.2.2	Django 是如何防范 CSRF 攻击的	207
	4.2.3	针对 CSRF 防御接口测试代码的调整	208
4.3	权限操作的漏洞		212
4.4	防止 XSS 攻击		218
4.5	防止 SQL 注入		218

参考文献 220

第1章

Python、Django 和 HTTP

Django 是基于 Python 语言的 Web 开发框架,所以要学习好 Django,首先要有基本的 Python 开发技巧,以及要了解 HTTP 的基本知识。本章首先介绍 Python 语言及其安装(Python 语法不在本书中介绍,读者可以查找其他书籍阅读),然后介绍 Django 知识及其安装,最后简单地介绍 HTTP。

Python 语言

1.1.1 Python 语言概述

在介绍 Python 之前,先来欣赏一下 Python 禅歌(读者可以在 Python 编译窗口中输入 import this 获得)。

> **英文原版**
>
> The Zen of Python,by Tim Peters
>
> Beautiful is better than ugly.
> Explicit is better than implicit.
> Simple is better than complex.
> Complex is better than complicated.
> Flat is better than nested.
> Sparse is better than dense.
> Readability counts.
> Special cases aren't special enough to break the rules.
> Although practicality beats purity.
> Errors should never pass silently.
> Unless explicitly silenced.
> In the face of ambiguity,refuse the temptation to guess.
> There should be one—and preferably only one—obvious way to do it.
> Although that way may not be obvious at first unless you're Dutch.

> Now is better than never.
> Although never is often better than *right* now.
> If the implementation is hard to explain, it's a bad idea.
> If the implementation is easy to explain, it may be a good idea.
> Namespaces are one honking great idea — let's do more of those!

> **中文翻译版及解释**
> 优美胜于丑陋
> 明了胜于晦涩
> 简洁胜于复杂
> 复杂胜于凌乱
> 扁平胜于嵌套
> 间隔胜于紧凑
> 可读性很重要
> 即便假借特例的实用性之名,也不可违背这些规则
> 不要包容所有错误,除非你确定需要这样做
> 当存在多种可能时,不要尝试去猜测
> 而是尽量找一种,最好是唯一一种明显的解决方案
> 虽然这并不容易,因为你不是 Python 之父
> 做也许好过不做,但不假思索就动手做还不如不做
> 如果你无法向人描述你的方案,那肯定不是一个好方案;反之亦然
> 命名空间是一种绝妙的理念,我们应当多加利用

Python 语言诞生于 20 世纪 90 年代初,它已被逐渐广泛地应用于系统管理任务的处理以及 Web 编程。Python 由于其易理解性、易读性以及简洁性,并且有近 30 年的历史,以及对云计算、大数据与人工智能(AI)开发(仅次于 R 语言)有很好的支持,因此越来越受到大众的喜爱。业界有种说法:"Java 十行代码,用 Python 一行代码就可以实现",这句话说得虽然有些夸张,但是"Java 三到四行代码,用 Python 一行代码就可以实现"是完全没有问题的。

Python 的创始人是 Guido van Rossum[①],在 1989 年圣诞节期间,住在阿姆斯特丹,为了打发圣诞节的无聊时光,决定开发一个新的脚本解释程序,作为 ABC 语言的一种继承。Guido 选用 Python(大蟒蛇的意思)作为该编程语言的名字,是因为他是一个名为 Monty Python 喜剧团体的爱好者。

ABC 语言是由 Guido 参加设计的一种教学语言。Guido 认为 ABC 语言非常优美以及

---

① Guido van Rossum(吉多·范·罗苏姆)1989 年在荷兰的国家数学和计算机科学研究院(Centrum voor Wiskunde en Informatica,CWI)创立了 Python 语言。1991 年初,Python 发布了第一个公开发行版。Guido 原居荷兰,1995 年移居美国,并遇到了他现在的妻子。2003 年年初,Guido 和他的家人,包括他 2001 年出生的儿子 Orlijn 一直居住在华盛顿州北弗吉尼亚郊区。随后他们搬迁到硅谷,2005 年开始他就职于 Google 公司,其中有一半时间是花在 Python 上,现在 Guido 在为 Dropbox 工作。——百度百科

非常强大，是专门为那些非专业程序员而设计的。但是，ABC语言并没有取得最后的成功，Guido认为主要是ABC语言的非开放性造成的，所以Guido决心在Python中避免这个错误。同时，他还想在ABC语言中实现他想过但是没有实现的东西。

Python在Guido手中诞生了，可以说，Python是从ABC语言发展起来的，主要受到Modula-3（另一种相当强大的语言，为小型团体所设计）的影响，并且结合了UNIX Shell和C的习惯。

Python已经成为最受欢迎的程序设计语言之一。2011年1月，它被TIOBE编程语言排行榜评为2010年度语言。2004年以后，Python的使用率呈线性增长。在2017年4月份TIOBE编程语言排行榜中，Python语言仅次于Java、C、C++和C#，位于第五位。

由于Python语言的简洁性、易读性以及可扩展性，在国外用Python做科学计算的研究机构日益增多，一些知名大学已经采用Python来教授程序设计课程。例如，卡耐基梅隆大学的编程基础、麻省理工学院的计算机科学及编程导论都是使用Python语言来讲授的。众多开源的科学计算软件包都提供了Python的调用接口。再如，著名的计算机视觉库OpenCV、三维可视化库VTK、医学图像处理库ITK。而Python专用的科学计算扩展库就更多了，例如，Java语言的三个十分经典的科学计算扩展库：NumPy、SciPy和matplotlib，它们分别为Python提供了快速数组处理、数值运算以及绘图功能。因此，Python语言及其众多的扩展库构成的开发环境十分适合工程技术、科研人员处理实验数据、制作图表，甚至开发科学计算应用程序。

1.1.2　Python的安装

目前，市场上呈现Python 2.x系列与Python 3.x系列共存的现象。读者可以安装Python 2.x系列或者Python 3.x系列。如果开发的目的是基于原有Python 2.x系列产品的维护，建议选择Python 2.x系列；如果目的是开发一个完全新的产品，那么建议选择Python 3.x系列。作者写这本书的时候，Python的最高版本是3.6，但是作者觉得Python 3.6还是不成熟，所以本书选择的版本是Python 3.5。

Python工具的官方下载地址是http://www.python.org/download。

Python下载完毕以后，务必要配置好环境变量（本书全部基于Windows开发环境进行介绍）。

图1-1是PYTHON_HOME变量的配置，变量值为安装Python的文件路径，在Python 3.5中默认为C:\Users\<Your_ID>\AppData\Local\Programs\Python\Python35\。

图1-1　PYTHON_HOME变量的配置

图1-2是在 PATH 中增加的两个参数，设置%PYTHON_HOME%是为了可以在任意路径下运行 Python 命令；设置%PYTHON_HOME%\Scripts\是为了可以在任意路径下运行%PYTHON_HOME%\Scripts\路径下的命令，如 pip 或 pip3。

%PYTHON_HOME%
%PYTHON_HOME%\Scripts\

图1-2　PATH 中的配置

1.2　Django 框架

1.2.1　Django 介绍

1. Django 概况

Django 项目是一个 Python 语言定制框架，它源自一个在线新闻 Web 站点，于 2005 年以开源的形式被释放出来。Django 框架的核心组件如下。

(1) 用于创建模型的对象关系映射。

(2) 为最终用户设计完美的管理界面。

(3) 一流的 URL 设计。

(4) 设计者友好的模板语言。

(5) 缓存系统。

Django 是用 Python 语言开发的一个开源的 Web 开发框架（Open Source Web Framework，OSWF），它鼓励快速开发，并遵循 MVC 设计理念。Django 遵守 BSD 版权[①]，初次发布于 2005 年 7 月，并于 2008 年 9 月发布了第一个正式版本 1.0。

Django 根据比利时的爵士音乐家 Django Reinhardt 命名，他是一个吉普赛人，主要以演奏吉它为主，还演奏过小提琴等。

由于 Django 在近年来的迅速发展，应用越来越广泛，被著名 IT 开发杂志 SD Times[②] 评选为 2013 SD Times 100，位列"API、库和框架"分类第六位，被认为是该领域的佼佼者。

2. Django 的设计理念

Django 的主要目的是简便、快速地开发数据库驱动的网站。它强调代码的复用以及多个组件可以很方便地以"插件"形式服务于整个框架。Django 有许多功能强大的第三方插件，甚至可以很方便地开发出自己的工具包，这使得 Django 具有很强的可扩展性。Django 还强调快速开发和 DRY(Do Not Repeat Yourself)的原则。

Django 基于 MVC 的设计十分优美。

(1) 对象关系映射(Object-Relational Mapping，ORM)：以 Python 类形式定义数据模型，ORM 将模型与关系数据库连接起来，得到一个非常容易使用的数据库 API。虽然在 Django 中可以使用原始的 SQL 语句，但一般从安全角度考虑，不建议这样做，因为一是 Django 已经对 SQL 语句进行了很好的封装；二是显示 SQL 语句容易引发类似 SQL 注入的威胁。本书将在第 2.8 节中详细介绍 ORM。

---

①　BSD（Berkeley Software Distribution，伯克利软件套件）是 UNIX 的衍生系统，1977—1995 年由加州大学伯克利分校开发和发布。——百度百科

②　*SD Times* 即《软件开发时代》杂志。

(2) URL分配：使用正则表达式匹配URL，就可以设计任意的URL。本书将在2.9.1节进行详细介绍。

(3) 模板系统：Django提供强大而可扩展的模板语言，它可以分隔设计、内容和Python代码，并且具有可继承性。本书将在2.10节进行详细介绍。

(4) 表单处理：可以方便地生成各种表单模型，实现表单的有效性检验。可以方便地从定义的模型实例生成相应的表单。本书将在3.3节进行详细介绍。

(5) Cache系统：可以挂在内存缓冲或其他的框架实现超级缓冲。

(6) 会话(session)：用户登录与权限检查，快速开发用户会话功能。本书将在2.6节进行详细介绍。

(7) 国际化：内置国际化系统，方便开发出多种语言的网站。

(8) 自动化的管理界面：不需要用大量的工作对后台内容进行维护。Django自带一个Admin Site，类似于后台管理系统。

3. 工作原理

(1) 用manage.py runserver启动Django服务器。

(2) 同时载入同一目录下的settings.py。该文件包含了项目中的配置信息，如URLConf等，其中最重要的配置就是ROOT_URLCONF，它告诉Django哪个Python模块应该用作本站的URLConf，如图1-3所示。

```
53  ROOT_URLCONF = 'ebusiness.urls'
```

图1-3 settings.py中的ROOT_URLCONF

(3) 当访问URL的时候，Django会根据ROOT_URLCONF的设置来装载URLConf。

(4) 然后按顺序逐个匹配URLConf里的URLpatterns。如果找到，则会调用相关联的视图方法，并把HttpRequest对象作为第一个参数(通常是request)。

(5) 最后，该view方法负责返回一个HttpResponse对象。

Django的工作原理如图1-4所示。

1.2.2　Django的安装

安装完毕Python，接下来安装Django，安装Django有以下4种方法。

1. 利用pip安装

由于1.1.2节中已经在PATH变量中增加了%PYTHON_HOME%\Scripts\项，所以可以在任意路径下运行如下命令。

```
>pip install django[==version]
```

[==version]可以不书写，不书写表示默认安装的是最新版本。

另外，卸载的方法是：

```
>pip uninstall django
```

2. 利用tar.gz安装

首先下载gz包，如Django-1.10.3.tar.gz文件，其中1.10.3是Django的版本号，然后

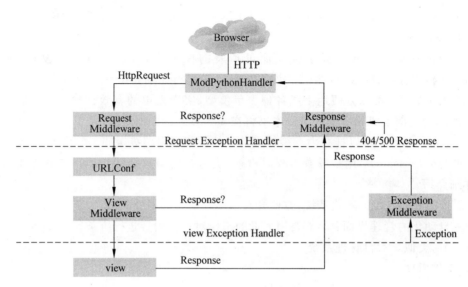

图 1-4　Django 的工作原理

进入目录内运行如下命令。

```
...>python setup.py install
```

3. 利用.whl 安装

wheel 文件是一个类似 zip 的文件包，其实用 pip 安装也是先安装 wheel 文件到本地，然后自动运行加压包的动作。首先下载 wheel 文件包，如 Django-1.10.3-py2.py3-none-ane-any.whl 文件，其中 1.10.3 仍旧是 Django 的版本号，然后运行如下命令。

```
...>pip install Django-1.10.3-py2.py3-none-ane-any.whl
```

4. 在 GitHub 上安装

可以利用类似于 Eclipse、Atom 到 GitHub 网站上安装 Djando。https://github.com/django/django 是 Djando 在 GitHub 上的地址。

1.3　HTTP 概述

超文本传输协议（HyperText Transfer Protocol，HTTP）是互联网上应用最为广泛的一种网络协议，所有的 3W 文件都必须遵守这个标准。设计 HTTP 最初的目的是为了提供一种发布和接收 HTML 页面的方法。1960 年，美国人 Ted Nelson 构思了一种通过计算机处理文本信息的方法，并称其为超文本（HyperText），这就是 HTTP 标准架构的发展根基，HTTP 的第一个版本 HTTP 0.9 是一种简单地用于网络间原始数据传输的协议。Ted Nelson 组织协调万维网协会（World Wide Web Consortium，WWW）和互联网工程工作小组（Internet Engineering Task Force，IETF）共同合作研究，最终发布了一系列的 RFC。HTTP 1.0 是在 RFC 1945 定义的，它在 HTTP 0.9 基础上做了改进，允许消息以类多用途因特网邮件扩展（Multipurpose Internet Mail Extensions，MIME）信息格式存在，包括请求/响应范式中的已传输和修饰符等方面的内容。现在普遍使用的是 RFC 2616 定义的 HTTP 1.1,要求严格保证可服务性,增强了在 HTTP 1.0 中没有考虑分层代理服务器、高速缓存、持久连接需求以及虚拟主机方面的能力。

现在 HTTP 还推出了 HTTP 2.0 版本。这里简单地介绍一下 HTTP 2.0。百度百科中对于 HTTP 2.0 是这样定义的:HTTP 2.0 即超文本传输协议 2.0,是下一代 HTTP,是由互联网工程任务组(IETF)的 HyperText Transfer Protocol Bis (httpbis)工作小组开发的,是自 1999 年 HTTP 1.1 发布后的首个更新。HTTP 2.0 在 2013 年 8 月进行首次合作共事性测试。在开放互联网上,HTTP 2.0 将只用于"https://网址",而"http://网址"将继续使用 HTTP 1.1,目的是在开放互联网上增加使用加密技术,以提供强有力的保护遏制主动攻击。DANE RFC6698 允许域名管理员不通过第三方 CA(认证授权)机构自行发行证书。

1.3.1 HTTP 的工作原理

HTTP 是基于 TCP 的,同时也可以承载 TLS 或 SSL 协议层,这里把承载 TLS 或 SSL 协议称作为 HTTPS。一般情况下,HTTP 为 80 端口,而 HTTPS 为 443 端口。图 1-5 是 HTTP 协议栈。图 1-6 是 HTTPS 协议栈。

图 1-5　HTTP 协议栈　　　　图 1-6　HTTPS 协议栈

通过图 1-7 可以更好地了解 HTTP 在整个网络中的位置。

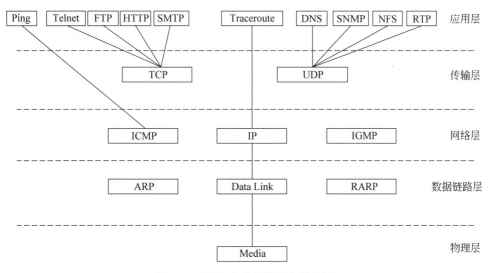

图 1-7　HTTP 在其他协议中的位置

1.3.2 HTTP 的请求

HTTP 的请求方式共分为 OPTIONS、GET、HEAD、POST、PUT、DELETE、TRACE 和 CONNECT 8 种（注意：这些方法均为大写），其中比较常用的为 GET 和 POST。

（1）OPTIONS：返回服务器针对特定资源所支持的 HTTP 请求方法，也可以利用向 Web 服务器发送 * 的请求来测试服务器的功能性。

（2）HEAD：向服务器所要与 GET 请求相一致的响应，只不过 HEAD 的请求响应体不会被返回（也就是说，GET 请求的响应为 HEAD 请求的响应＋响应体）。这种方法可以在不必传输整个响应内容的情况下，就可以获取包含在响应小消息头中的元信息。

（3）GET：向特定的资源发出请求。注意：GET 方法不应当被用于产生"副作用"的操作中。例如，在 Web Application 中，其中一个原因是 GET 可能会被网络蜘蛛等随意访问。

（4）POST：向指定资源提交数据处理请求（如提交表单或者上传文件）。数据被包含在请求体中。POST 请求可能会导致新资源的建立和（或）已有资源的修改。

（5）PUT：向指定资源位置上传其最新内容。

（6）DELETE：请求服务器删除 Request-URL 所标识的资源。

（7）TRACE：回显服务器收到的请求，主要用于测试或诊断。

（8）CONNECT：HTTP 1.1 协议中预留给能够将连接改为管道方式的代理服务器。

HTTP 的请求分为以下三部分。

① 请求行。

② 请求头。

③ 请求正文。

图 1-8 是一个用 Fiddler 4 捕捉到的访问 http://www.3testing.com 网站的请求内容。

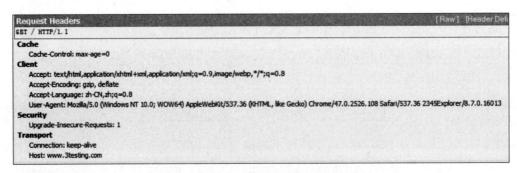

图 1-8 Fiddler 4 捕捉 HTTP 的请求信息

其中，GET / HTTP/1.1 为请求行，GET 表示请求方法，包括前面介绍的 8 种方法之一；/表示访问的是根目录；HTTP/1.1 表示协议版本号为 1.1。

后面的都为请求头，关于请求头的具体介绍，读者可以上 RFC 2616 官方网站查询，这里不进行介绍了。

由于这个请求没有请求数据，所以没有请求正文。图 1-9 是一个 Fiddler 4 捕捉 HTTP 的请求正文例子。

```
------WebKitFormBoundarybWudMLpxmT9YA752
Content-Disposition: form-data; name="csrfmiddlewaretoken"

ljRgbiYPTVkyYjxeYC9bWMNW3RljmvNDPp740LYHp2XOOWOJgEe8WbCqdAwOKGal
------WebKitFormBoundarybWudMLpxmT9YA752
Content-Disposition: form-data; name="username"

cindy
------WebKitFormBoundarybWudMLpxmT9YA752
Content-Disposition: form-data; name="password"

123456
```

图 1-9 Fiddler 4 捕捉 HTTP 的请求正文例子

1.3.3　HTTP 的应答

HTTP 的应答返回码包含服务器响应情况，见表 1-1。

表 1-1　HTTP 的应答返回码

消　　息	描　　述
100 Continue	客户应该和自己的请求继续。中间的应答被用于告知客户请求的初始部分已经收到，并且还没有被服务器拒绝
101 Switching Protocols	服务器转换协议：服务器将遵从客户的请求转换到另外一种协议
200 OK	请求成功（其后是对 GET 和 POST 请求的应答文档）
201 Created	请求被创建完成，同时新的资源被创建
202 Accepted	提供处理的请求已被接受，但是处理未完成
203 Non-authoritative Information	文档已经正常返回，但一些应答头可能不正确，因为使用的是文档的副本
204 No Content	没有新文档。浏览器应该继续显示原来的文档。如果用户定期刷新页面，而 Servlet 可以确定用户文档足够新，这个状态代码就是有用的
205 Reset Content	没有新文档。但浏览器应该重置它所显示的内容，用来强制浏览器清除表单中输入的内容
206 Partial Content	客户发送了一个带有 Range 头的 GET 请求，服务器响应了该请求
300 Multiple Choices	多重选择。链接列表。用户可以选择某链接到达目的地。最多允许五个地址
301 Moved Permanently	请求的页面已经转移至新的 URL
302 Found	请求的页面已经临时转移至新的 URL
303 See Other	请求的页面可在别的 URL 下被找到
304 Not Modified	未按预期修改文档。客户端有缓冲的文档并发出了一个条件性的请求（一般是提供 If-Modified-Since 头表示客户只想比指定日期更新的文档）。服务器告诉客户，原来缓冲的文档还可以继续使用

续表

消息	描述
305 Use Proxy	客户请求的文档应该通过 Location 头所指明的代理服务器提取
306 Unused	此代码用于前一版本,目前已不再使用,但是代码依然被保留
307 Temporary Redirect	被请求的页面已经临时移至新的 URL
400 Bad Request	错误的请求
401 Unauthorized	被请求的页面需要用户名和密码
401.1	登录失败
401.2	服务器配置导致登录失败
401.3	由于 ACL(访问控制列表)对资源的限制而未获得授权
401.4	筛选授权失败
401.5	ISAPI/CGI(即 Internet 服务的应用程序接口/通用网关接口)应用程序授权失败
401.7	访问被 Web 服务器上的 URL 授权策略拒绝。IIS 6.0 专用代码
402 Payment Required	尚无法使用
403 Forbidden	被禁止请求页面的访问
403.1	被禁止执行访问
403.2	被禁止读访问
403.3	被禁止写访问
403.4	要求 SSL
403.5	要求 SSL 128
403.6	被拒绝 IP 地址
403.7	要求客户端证书
403.8	被拒绝站点访问
403.9	用户数过多
403.10	配置无效
403.11	密码更改
403.12	映射表访问被拒绝
403.13	客户端证书被吊销
403.14	拒绝目录列表
403.15	超出客户端访问许可
403.16	客户端证书不受信任或无效
403.17	客户端证书已过期或尚未生效

续表

消　息	描　述
403.18	在当前的应用程序池中不能执行所请求的URL。IIS 6.0专用代码
403.19	不能为这个应用程序池中的客户端执行CGI。IIS 6.0专用代码
403.20	PASSPORT登录失败。IIS 6.0专用代码
404 Not Found	服务器无法找到被请求的页面
404.0	没有找到文件或目录
404.1	无法在所请求的端口上访问Web站点
404.2	Web服务扩展锁定策略阻止本请求
404.3	MIME映射策略阻止本请求
405 Method Not Allowed	不被允许请求中指定的方法
406 Not Acceptable	客户端接受了浏览器不能解释的返回消息
407 Proxy Authentication Required	用户必须首先使用代理服务器进行验证,这样请求才可以被处理
408 Request Timeout	请求超出了服务器的等待时间
409 Conflict	由于冲突,请求无法被完成
410 Gone	被请求的页面不可用
411 Length Required	"Content-Length"未被定义。如果没有这个内容,服务器就不会接受请求
412 Precondition Failed	服务器评估请求中的前提条件为失败
413 Request Entity Too Large	由于请求的实体太大,所以服务器不会接受请求
414 Request-url Too Long	由于URL太长,服务器不会接受请求。当POST的请求被转换为GET请求的时候,就会触发这个情况
415 Unsupported Media Type	由于媒介类型不被支持,所以服务器不会接受请求
416 Requested Range Not Satisfiable	服务器不能满足客户在请求中指定的Range头
417 Expectation Failed	执行失败
423	锁定的错误
500 Internal Server Error	请求未完成。服务器遇到不可预知的情况
500.12	应用程序正忙于在Web服务器上重新启动
500.13	Web服务器太忙
500.15	不允许直接请求Global.asa
500.16	UNC授权凭据不正确。IIS 6.0专用代码
500.18	URL授权存储不能打开。IIS 6.0专用代码
500.100	内部ASP错误

续表

消　　息	描　　述
501 Not Implemented	请求未完成。服务器不支持所请求的功能
502 Bad Gateway	请求未完成。服务器从上游服务器收到一个无效的响应
502.1	CGI 应用程序超时
502.2	CGI 应用程序出错
503 Service Unavailable	请求未完成。服务器临时过载或死机
504 Gateway Timeout	网关超时
505 HTTP Version Not Supported	服务器不支持请求中指明的 HTTP 版本

上述的返回码共分为以下 5 类。

(1) 1XX:指示信息,表示接收到请求,继续进程。

(2) 2XX:成功,表示请求已被成功接收、理解和接受。

(3) 3XX:重定向,要完成请求必须进行更进一步的操作。

(4) 4XX:客户端错误,请求有语法错误或者无法实现。

(5) 5XX:服务器错误,服务器未能实现合法请求。

HTTP 的应答与请求非常相似,也分为以下 3 部分。

(1) 应答行。

(2) 应答头。

(3) 应答正文。

图 1-10 是用 Fiddler 4 捕捉到访问 http://www.3testing.com 网站的应答内容。

图 1-10　用 Fiddler 4 捕捉到访问 http://www.3testing.com 网站的应答内容

其中,HTTP/1.1 200 OK 为应答行,如 1.3.2 节一样,HTTP/1.1 表示 HTTP 版本编号;200 表示返回码,包括前面提到五类中的任意一个;OK 表示返回短语。

下面的都为应答头,读者也可以上 RFC 2616 官方网站查询。

返回正文就是通常被看到的 HTML 代码。

1.3.4 HTTP 的连接性

通信中无连接的含义是限制每次连接只处理一个请求。服务器处理完客户端的请求，并收到客户的应答后，就断开连接。采用这种方式可以节省传输时间。在日常生活中可以认为普通邮件（是 Mail，非 Email）是无连接的，而打电话是有连接的。当发送邮件的时候，虽然信封上有收件人的地址和邮编，但是邮件有无收到若不通过其他方式是不可能知道的，所以无连接的通信是不可靠的；而打电话是有连接的，正常情况包括拨号、应答和挂断，如果对方正在通话，则显示忙音；如果对方不在现场，则显示无人应答，所以有连接的通信是可靠的。

HTTP 是无连接的，这是由于早期 HTTP 产生的时候，服务器需要同时处理面向全世界数十万，甚至上百万个客户端的网页访问，但是每个浏览器与服务器之间交换的间歇性是比较大的，并且网页浏览的发散性导致两次传送的数据关联性很低，大部分的通道实际上会很空闲、无端占用资源，所以，HTTP 的设计者有意利用这种特点将协议设计为请求时创建连接、请求完释放连接，即面向无连接的，以尽快将资源释放出来服务给其他客户端。

但是，随着时间的推移，网页变得越来越复杂，网页里存在很多图片、视频等文件，这种情况下如果每次访问都需要重新建立一次 TCP 连接就显得非常低效了。因此，Keep-Alive 在 HTTP 1.1 中被提出用来解决这个问题。

Keep-Alive 可以使客户端到服务器端的连接持续有效，当出现对服务器的后续请求时，Keep-Alive 能够避免建立或者重新建立连接。大部分 Web 服务器，包括 Django、IIS 和 Apache，都支持 HTTP Keep-Alive。对于提供静态内容的网站来说，这个功能通常是非常有用的。但是，对于负担较重的网站来说，这里存在另外一个问题，即对性能的影响。当 Web 服务器和应用服务器在同一台机器上运行时，Keep-Alive 功能对资源利用的影响尤其突出。

有了 Keep-Alive，客户端和服务器之间的 HTTP 连接就会被保持，不会断开，当客户端发送另外一个请求时，就使用这条已经建立的连接。

图 1-11 是基于 HTTP 1.0 的页面请求。

(1) 浏览器与 Web 服务器建立连接。
(2) 浏览器向 Web 服务器发送 HTTP 网页 1 的请求。
(3) Web 服务器向浏览器返回网页 1 的响应消息。
(4) 浏览器与 Web 服务器断开连接。
(5) 浏览器与 Web 服务器建立连接。
(6) 浏览器向 Web 服务器发送图片 1.1 请求。
(7) Web 服务器向浏览器返回图片 1.1 的响应消息。
(8) 浏览器与 Web 服务器断开连接。

……

(n) 浏览器与 Web 服务器建立连接。
($n+1$) 浏览器向 Web 服务器发送 HTTP 网页 2 的请求。
($n+2$) Web 服务器向浏览器返回网页 2 的响应消息。
($n+3$) 浏览器与 Web 服务器断开连接。

($n+4$)浏览器与 Web 服务器建立连接。
($n+5$)浏览器向 Web 服务器发送图片 2.1 请求。
($n+6$)Web 服务器向浏览器返回图片 2.1 的响应消息。
($n+7$)浏览器与 Web 服务器断开连接。
……

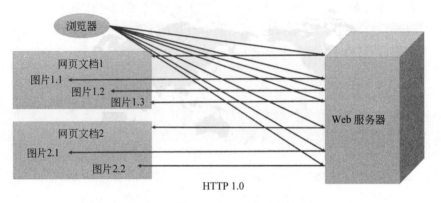

图 1-11　基于 HTTP 1.0 的页面请求

图 1-12 是基于 HTTP 1.1 的页面请求，这里加入了 HTTP Keep-Alive。

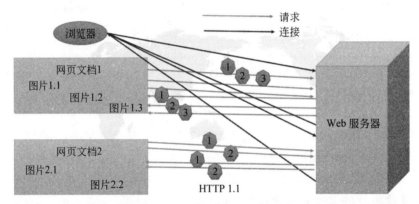

图 1-12　基于 HTTP 1.1 的页面请求

(1)浏览器与 Web 服务器建立连接。
(2)浏览器向 Web 服务器发送 HTTP 网页 1 的请求。
(3)Web 服务器向浏览器返回网页 1 的响应消息。
(4)浏览器向 Web 服务器发送图片 1.1 请求。
(5)浏览器向 Web 服务器发送图片 1.2 请求。
(6)浏览器向 Web 服务器发送图片 1.3 请求。
(7)Web 服务器向浏览器返回图片 1.1 的响应消息。
(8)Web 服务器向浏览器返回图片 1.2 的响应消息。
(9)Web 服务器向浏览器返回图片 1.3 的响应消息。
(10)浏览器与 Web 服务器断开连接(注意：不同的 HTML 页面不能通过 HTTP Keep-Alive 保持连接)。

（11）浏览器与 Web 服务器建立连接。
（12）浏览器向 Web 服务器发送 HTTP 网页 2 的请求。
（13）Web 服务器向浏览器返回网页 2 的响应消息。
（14）浏览器向 Web 服务器发送图片 2.1 请求。
（15）浏览器向 Web 服务器发送图片 2.2 请求。
（16）Web 服务器向浏览器返回图片 2.1 的响应消息。
（17）Web 服务器向浏览器返回图片 2.2 的响应消息。
（18）浏览器与 Web 服务器断开连接。

1.3.5　HTTP 的无状态

通信中无状态协议是指同一个会话的连续两个请求互相不了解，它们由最新实例化的环境进行解析，除了应用本身可能已经存储在全局对象中的所有信息外，该环境不保存与会话有关的任何信息。

HTTP 是一个无状态协议，这意味着每个请求都是独立的，Keep-Alive 不能改变这个结果。

缺少状态意味着如果后续处理需要前面的信息，则必须重传，这样可能导致每次连接传送的数据量增大。另一方面，在服务器不需要先前信息时，它的应答就较快。

HTTP 这种特性既有优点，也有缺点，优点是服务器得到了解放，每次请求"点到为止"不会造成不必要的连接占用，缺点是每次请求会传输大量重复的内容信息。

进行动态交互的 Web 应用程序出现之后，HTTP 无状态的性质严重阻碍了这些应用程序的实现，这是因为交互是需要承前启后的，例如，"购物车"的程序就要知道用户到底在之前选择了什么商品。这样，两种用于保持 HTTP 连接状态的技术就应运而生了，它们分别是 cookie 和 session。

cookie 可以保持登录信息到用户下次与服务器的会话，用户这次登录后，下次登录就不需要输入用户名和密码了。还有一些 cookie 在用户退出会话的时候就被删除了，这样可以有效保护个人隐私（这种 cookie 叫作非持久型 cookie，具有固定会话期限的 cookie 叫作持久型 cookie）。

cookie 最典型的应用是判定注册用户是否已经登录了网站，用户可能会被提示是否在下一次进入此网站时保留用户信息，以便简化登录手续，这些都是 cookie 的功用。另一个重要的应用场合是"购物车"之类的处理。用户可能会在一段时间内在同一家网站的不同页面中选择不同的商品，这些信息都会写入 cookie，以便在最后付款时提取信息。图 1-13 为网易 126 邮箱网站通过 cookie 保持登录选项。图 1-14 为京东网站中购物车中内容的显示。

与 cookie 类似的另外一个解决方案是 session，它是通过服务器来保持状态的。

当客户端访问服务器的时候，服务器会根据需求设置 session，将会话信息保存在服务器上，同时将标示 session 的 sessionid 传送给客户端浏览器，浏览器将这个 sessionid 保存在客户端的内存中，通常称 sessionid 为没有过期时间的 cookie。当浏览器关闭后，这个 cookie 就会被清掉，它不会存在于用户的 cookie 临时文件。

以后浏览器每次请求都会额外加上这个参数值，服务器根据这个 sessionid 就能取得客

图 1-13　网易 126 邮箱网站通过 cookie 保持登录选项

图 1-14　京东网站中购物车中内容的显示

户端的数据信息。

如果客户端浏览器意外关闭，服务器保存的 session 数据是不会立即释放的，这个时候数据还会存在，只要知道那个 sessionid，就可以继续通过请求获得此 session 的信息，因为此时后台的 session 还存在，当然可以设置一个 session 超时时间，一旦超过规定时间没有客户端请求，服务器就会清除对应 sessionid 的 session 信息。

第 2 章

Django 基本知识

由于 Django 版本更新迭代太快,所以有些命令每个版本不太一样,本书是基于 1.11.4 版本编写的。查看 Django 版本的方法有 3 种。

(1) 打开 Python 解释器。
(2) >>>import django。
(3) >>>print (django.VERSION)。

图 2-1 是作者机器上的 Django 的版本输出,表示当前 Django 版本是 1.11.4。其中 final 表示最终版本,如果是 dev,就表示开发版本。

图 2-1 Django 的版本输出

2.1 启动 Django 服务

要打开网站,必须先启动网站服务器。安装完 Django 后,在％PYTHON_HOME％\Scripts\目录下看见有 django-admin.exe 和 django-admin.py 两个文件。打开 Windows 命令框,进入项目工作目录,在这里把工作目录建立在％PYTHON_HOME％\Scripts\下,之后运行命令 scripts\>django-admin。

然后会有如下这些参数提示。

```
Type 'django-admin help <subcommand>' for help on a specific subcommand.

Available subcommands:

[django]
    check
    compilemessages
    createcachetable
    dbshell
    diffsettings
    dumpdata
    flush
    inspectdb
    loaddata
```

```
            makemessages
            makemigrations
            migrate
            runserver
            sendtestemail
            shell
            showmigrations
            sqlflush
            sqlmigrate
            sqlsequencereset
            squashmigrations
            startapp
            startproject
            test
            testserver
Note that only Django core commands are listed as settings are not properly
configured (error: Requested setting INSTALLED_APPS, but settings are not
configured. You must either define the environment variable DJANGO_SETTINGS_MODULE
or call settings.configure() before accessing settings.).
```

这里列出的都是 django-admin 命令参数的用法,现在用参数 startproject 来建立一个名字为 ebusiness 的项目。

```
\scripts>django-admin startproject ebusiness
```

运行这个命令后,会看到在当前目录下新建了一个名为 ebusiness 的目录,其结构及解释如下。

ebusiness

|---------- ebusiness

| |----------__int__.py:空文件,表示目录为 Python 的标准包。

| |----------settings.py:Django 的配置文件。

| |----------urls.py:Django 项目的 URL 文件,这个文件会经常用到。

| |----------wsgi.py:与 WSGI[①] 兼容的 Web 服务器,为项目提供服务的入口点。

|----------mange.py:命令行工具,用来同步数据库、启动服务等。

接下来通过\scripts>cd ebusiness 进入 ebusiness 目录,运行\ebusiness>python manage.py 命令,会有如下这些参数提示。

```
Type 'manage.py help <subcommand>' for help on a specific subcommand.
```

[①] WSGI 是 Web Server Gateway Interface 的缩写。从层的角度来看,WSGI 所在层的位置低于 CGI。但与 CGI 不同的是,WSGI 具有很强的伸缩性且能运行于多线程或多进程的环境下,这是因为 WSGI 只是一份标准,并没有定义如何去实现。实际上,WSGI 并非 CGI,因为其位于 Web 应用程序与 Web 服务器之间,而 Web 服务器可以是 CGI、mod_python(注:现通常使用 mod_wsgi 代替)、FastCGI 或者是一个定义了 WSGI 标准的 Web 服务器,就像 Python 标准库提供的独立 WSGI 服务器称为 wsgiref。——百度百科

```
Available subcommands:

[auth]
    changepassword
    createsuperuser

[contenttypes]
    remove_stale_contenttypes

[django]
    check
    compilemessages
    createcachetable
    dbshell
    diffsettings
    dumpdata
    flush
    inspectdb
    loaddata
    makemessages
    makemigrations
    migrate
    sendtestemail
    shell
    showmigrations
    sqlflush
    sqlmigrate
    sqlsequencereset
    squashmigrations
    startapp
    startproject
    test
    testserver

[sessions]
    clearsessions

[staticfiles]
    collectstatic
    findstatic
    runserver
```

上面列出的都是 manage.py 命令参数的用法,在这里用参数 startapp 来建立一个应用,如\ebusiness>python manage.py startapp goods。

接下来可以看到在当前目录下多了一个 goods 目录(我们称 goods 为一个应用,一个项目下面可以建立一到多个应用)。

goods

|----------- migrations/:记录数据库模型中的数据变更。

|----------- admin. py:映射数据库模型中的数据到 Django 自带的后台服务。

|----------- apps. py:用于应用程序的配置。

|----------- models. py:Django 数据库模型文件。

|----------- tests. py:创建 Django 测试用例。

|----------- views. py:Django 视图文件,用于向前端页面传输内容,程序中的多数逻辑处理都在这里,使用也相当频繁。

这时通过命令\ebusiness>python manage. py runserver 就可以运行项目了。

结果如图 2-2 所示。

```
C:\Python34\Scripts\ebusiness>python manage.py runserver
Performing system checks...

System check identified no issues (0 silenced).
August 23, 2017 - 16:25:09
Django version 1.11.4, using settings 'ebusiness.settings'
Starting development server at http://127.0.0.1:8000/
Quit the server with CTRL-BREAK.
```

图 2-2 启动 Web Server

打开浏览器,输入 http://127. 0. 0. 1:8000,可以看到服务器已经出现了,但是由于目前没有进行任何代码的编写,所以还看不到任何内容(注意,Django 的默认端口为 8000)。

另外,对于 python manage. py 后面参数的具体使用方法,可以通过命令 manage. py help <命令选项>来实现,例如/ebusiness>manage. py help runserver 会显示 runserver 如何使用,具体如图 2-3 所示。

```
C:\Python34\Scripts\ebusiness>manage.py help runserver
usage: manage.py runserver [-h] [--version] [-v {0,1,2,3}]
                           [--settings SETTINGS] [--pythonpath PYTHONPATH]
                           [--traceback] [--no-color] [--ipv6] [--nothreading]
                           [--noreload] [--nostatic] [--insecure]
                           [addrport]

Starts a lightweight Web server for development and also serves static files.

positional arguments:
  addrport              Optional port number, or ipaddr:port

optional arguments:
  -h, --help            show this help message and exit
  --version             show program's version number and exit
  -v {0,1,2,3}, --verbosity {0,1,2,3}
                        Verbosity level; 0=minimal output, 1=normal output,
                        2=verbose output, 3=very verbose output
  --settings SETTINGS   The Python path to a settings module, e.g.
                        "myproject.settings.main". If this isn't provided, the
                        DJANGO_SETTINGS_MODULE environment variable will be
                        used.
  --pythonpath PYTHONPATH
                        A directory to add to the Python path, e.g.
                        "/home/djangoprojects/myproject".
  --traceback           Raise on CommandError exceptions
  --no-color            Don't colorize the command output.
  --ipv6, -6            Tells Django to use an IPv6 address.
  --nothreading         Tells Django to NOT use threading.
  --noreload            Tells Django to NOT use the auto-reloader.
  --nostatic            Tells Django to NOT automatically serve static files
                        at STATIC_URL.
  --insecure            Allows serving static files even if DEBUG is False.
```

图 2-3 manage. py help runserver

2.2 Hello World 程序

Hello World 是每个程序开发者在初学一门语言框架时必须做的一个项目,这里也先来开发这样一个应用。

2.2.1 直接打印显示内容

首先在 ebusiness/setting.py 文件中加入 goods 应用,见下面粗体字内容。

```
#Application definition

INSTALLED_APPS = [
    'django.contrib.admin',
    'django.contrib.auth',
    'django.contrib.contenttypes',
    'django.contrib.sessions',
    'django.contrib.messages',
    'django.contrib.staticfiles',
    'goods'
]
```

然后在 ebusiness/urls.py 中输入以下粗体字内容。

```
...
from django.conf.urls import url
from django.contrib import admin
from goods import views #导入 goods 应用 views 文件
...
urlpatterns = [
    url(r'^admin/', admin.site.urls),
    url(r'^index/$', views.index),
]
```

最后打开 goods/views.py 文件,输入以下粗体字内容。

```
...
#coding=utf-8
from django.shortcuts import render
from django.http import HttpResponse
...

def index(request):
    return HttpResponse("Hello World!")
```

这里,index(request)被称为基于函数的视图,这个函数称为"视图函数"。另外,在这种方式下,参数 request 是必不可少的,后面还可以跟其他参数。

这时打开浏览器输入 http://127.0.0.1:8000/index/,就可以看到字符串"Hello World!"了。

2.2.2 通过文件模板显示内容

除了用这个办法来显示网页外,通常使用比较多的方式是利用网页模板来显示。修改 goods/views.py 文件,修改的内容如以下粗体字部分。

```
...
#coding=utf-8
from django.shortcuts import render
from django.http import HttpResponse
...

def index(request):
    return render (request, "index.html")
```

然后在当前 goods 目录下建立一个 templates 文件夹(注意,目录名称一定为 templates,这是 Django 的规定),进入 templates 目录建立 index.html 文件模板。

```
<html>
<head>
</head>
<body>
<h1>Hello World!</h1>
</body>
</html>
```

刷新网页,就能够看见模板定义的 Hello World 界面了。

2.2.3 文件模板参数

下面介绍如何通过 views.py 中的参数传递给 HTML 模板的方法。

```
...
#coding=utf-8
from django.shortcuts import render, render_to_response
from django.http import HttpResponse
...

def index(request):
    return render_to_response('index.html',{"String":"Hello World!"})
```

可以看见,这里把 return render(request,"index.html")换成了 return render_to_response('index.html',{"String":"Hello World!"}),在方法 render_to_response()中,第

一个参数为模板的名字,第二个参数为一个字典类型,字典的键为传过去的参数,值为参数对应的值。然后再来改造 index.html 模板。

```
<html>
<head>
</head>
<body>
<h1>{{ String }}</h1>
</body>
</html>
```

可见,{{}}为传过来的参数,{{}}内的值为 render_to_response()方法第二个变量中字典中的参数,返回给网页的是参数对应的值。

关于模板的更多知识,本书将在 2.10 节中进行详细介绍。

2.3 获取参数

上面提到在 Web 开发中经常用到的 HTTP 传输方式为 GET 和 POST 两种方式,下面来讨论 Django 是如何获得上一个页面传输下来的 GET 和 POST 参数方法的。

2.3.1 通过 GET 方式获取

GET 方式比较简单,例如,一个 HTTP 请求的命令结尾为 ?id=1&username=jerry,可以通过如下代码实现。

```
...
id=request.GET.get("id", "")
name=request.GET.get("username ", "")
...
```

这里获得 id 的值为"1",name 的值为"jerry"。所以,Django 获得 GET 方法的代码为

变量 =request.GET.get("var", "")

其中 var 为 GET 的参数。

2.3.2 通过 POST 方式获取

1. 通过模板 POST 方式获得

POST 方式传输数据也是 Web 开发经常使用的。例如,有这么一个表单请求:

```
...
<form action="/login/" method="POST">
<input type="text" name="username">
<input type="password" name="password">
```

```
<input type="submit" value="登录">
</form>
...
```

可以通过如下方式获得 username 和 password 参数。

```
...
username=request.POST.get("username","")
password=request.POST.get("password","")
...
```

因此,Django 获得 POST 方法的代码为

```
变量=request.POST.get("var","")
```

其中 var 为 POST 参数。

2. 通过自定义 POST 方式获得

Django 可以自定义表单,首先介绍如何自定义一个表单。这里把这个表单定义在一个名为 forms.py 的文件中。

```python
...
from django import forms
...
#定义注册表单模型
class UserForm(forms.Form):
    username=forms.CharField(label='用户名',max_length=100)
    password=forms.CharField(label='密码',widget=forms.PasswordInput())
    email=forms.EmailField(label='电子邮件')
...
```

这里建立了三个表单元素,其中"用户名"和"密码"为文本格式 CharField,"用户名"中的 max_length=100 表示文本框的最大长度为 100;"密码"中通过 widget=forms.PasswordInput()表示以 type="password"形式显示;"电子邮件"为 EmailField,表示以 HTML5 的 Email 格式 type="email"形式来显示。这个表单最后生成的 HTML 文本为

```html
...
<p><label for="id_username">用户名:</label><input type="text" name="username" id="id_username" maxlength="100" required /></p>
<p><label for="id_password">密码:</label><input type="password" name="password" id="id_password" required /></p>
<p><label for="id_email">电子邮件:</label><input type="email" name="email" id="id_email" required /></p>
...
```

然后介绍如何获取参数。

```
...
#用户注册
def register(request):
    if request.method =="POST":
        uf =UserForm(request.POST)
        ...
    else:
        uf =UserForm()
    return render_to_response('register.html',{'uf':uf})
```

如果网页通过 POST 提交,则开始获得表单数据。

```
if request.method =="POST":
        uf =UserForm(request.POST)
```

否则,进入提交页面,显示表单内容。

```
else:
uf =UserForm()
    return render_to_response('register.html',{'uf':uf})
```

最后介绍如何通过这种方式获取表单信息。

```
...
#用户注册
def register(request):
    if request.method =="POST":
        uf =UserForm(request.POST)
        if uf.is_valid():
            #获取表单信息
            username =uf.cleaned_data['username']
            password =uf.cleaned_data['password']
            email =uf.cleaned_data['email']
```

所以,自定义表单的 POST 方式获取方式为

```
变量 =uf.cleaned_data['var']
```

这里的 var 同样为表单参数。

注意:表单数据获取 uf.cleaned_data 一定要在 if request.method == "POST":内,并且前面加上 if uf.is_valid():来判断输入的数据是否有误。

注册代码如下。

```
...
#用户注册
def register(request):
    if request.method =="POST":
        uf =UserForm(request.POST)
```

```
            if uf.is_valid():
                #获取表单信息
                username = uf.cleaned_data['username']
                password = uf.cleaned_data['password']
                email = uf.cleaned_data['email']
                #查找数据库中是否存在相同的用户名
                ...
                if user_list:
                    return render_to_response('register.html',{'uf':uf,"error":"用户名已经存在!"})
                else:
                    #将表单写入数据库
                    ...
                    #返回注册成功页面
                    uf = LoginForm()
                    return render_to_response('index.html',{'uf':uf})
            else:
                uf = UserForm()
                return render_to_response('register.html',{'uf':uf})
    ...
```

　　里面关于数据库的知识先隐藏起来，在后面章节中会进行详细介绍。当然，密码需要通过 MD5 加密方式进行存储，本书将在第 4 章中详细介绍。

　　这里需要提出，Django 默认对 CSRF 攻击通过调用 CSRF 令牌插件的方式进行了防范，在调用插件后每个 form 里面应该包括一个 CSRF 令牌，设置的方法是在模板文件中加入{{csrf_token}}。如果不使用 CSRF 令牌，可以在 setting.py 中将其注释掉。

```
    ...
    IDDLEWARE = [
        'django.middleware.security.SecurityMiddleware',
        'django.contrib.sessions.middleware.SessionMiddleware',
        'django.middleware.common.CommonMiddleware',
        #'django.middleware.csrf.CsrfViewMiddleware',
        'django.contrib.auth.middleware.AuthenticationMiddleware',
        'django.contrib.messages.middleware.MessageMiddleware',
        'django.middleware.clickjacking.XFrameOptionsMiddleware',
    ]
    ...
```

　　这里先把这个插件注释掉，本书 4.2 节会打开这个插件，详细介绍如何使用 CSRF 攻击防范。这里把 LoginForm 也定义在 forms.py 文件中。

```
    ...
    #定义登录表单模型
```

```
class LoginForm(forms.Form):
    username = forms.CharField(label='用户名',max_length=100)
    password = forms.CharField(label='密码',widget=forms.PasswordInput())
...
```

LoginForm 与 UserForm 的区别主要在于 UserForm 中没有 Email 域。

2.4 HttpRequest 对象与 HttpResponse 对象

2.4.1 HttpRequest 对象

当客户端向服务器发送请求的时候,Django 会创建一个名为 HttpRequest 的对象,并且通过参数 request 传给视图函数。下面对它的属性进行详细介绍。

(1) path:请求页面的全路径,但是不包括域名,如"/static/css/"。

(2) method:请求中使用 HTTP 方法的字符串表示,全大写表示。

例如,

```
if request.method =='GET':
    #dosomething()
elif request.method =='POST':
    #dootherthing()
```

① GET:包含所有 HTTP GET 参数的类字典对象。

② POST:包含所有 HTTP POST 参数的类字典对象。

注意:POST 不包括 file-upload 信息,参见 FILES 属性。

(3) REQUEST:该属性是 POST 和 GET 属性的集合体,但是有特殊性,先查找 POST 属性,然后再查找 GET 属性。

例如,如果 GET = {"name":"Jerry"},POST = {"age":"43"},则 REQUEST["name"]的值是"Jerry", REQUEST["age"]的值是"43"。

但是,还是强烈建议不要使用 **REQUEST**,而是使用 **GET** 和 **POST**,因为这两个属性更加显式化,写出的代码也更易被理解。

(4) COOKIES:包含所有 cookie 的标准 Python 字典对象。Keys 和 values 都是字符串。

(5) FILES:包含所有上传文件的类字典对象。FILES 中的每个 Key 都是<input type="file" name="" />标签中 name 属性的值。FILES 中的每个 value 同时也是一个标准 Python 字典对象,包含下面三个 Keys。

① filename:上传的文件名,用 Python 字符串表示。

② content-type:上传文件的内容类型。

③ content:上传文件的原始内容。

注意:只有在请求方法是 POST,并且请求页面中<form>有 enctype="multipart/

form-data"属性时,FILES才拥有数据,否则,FILES就是一个空的字典。

(6) META:字典类型,包含所有可用的 HTTP 头部信息。例如,CONTENT_LENGTH、CONTENT_TYPE、QUERY_STRING(未解析的原始查询字符串)、REMOTE_ADDR(客户端 IP 地址)、REMOTE_HOST(客户端主机名)、SERVER_NAME(服务器主机名)、SERVER_PORT(服务器端口)。

(7) User:是一个 django.contrib.auth.models.User 的对象,表示当前登录的用户。如果用户当前没有登录,user 将被初始化为 django.contrib.auth.models.AnonymousUser 的实例。可以通过 user 的 is_authenticated()方法来辨别用户是否已登录。

```
if request.user.is_authenticated():
    #Do something for logged-in users.
else:
    #Do something for other users.
```

注意:只有激活 Django 中的 AuthenticationMiddleware 时,该属性才可以使用。

(8) session:这是一个唯一可以读写的属性,表示当前会话的字典对象。注意,只有激活 Django 中的 session 支持时,该属性才可用。

(9) raw_post_data:原始 HTTP POST 数据,没有解析过,它只有在高级处理的时候时才会有用处。

2.4.2 HttpResponse 对象

对于 HttpRequest 对象,是由 Django 自动创建,而对于 HttpResponse 对象,就必须由用户自己创建了。每个 view 方法必须返回一个 HttpResponse 对象。对于 HttpResponse 使用方法,2.2.1 节中已进行了介绍。HttpResponse 对象下有一些子对象,下面对其进行详细介绍。

(1) HttpResponseRedirect:构造函数接受单个参数,重定向到的 URL。可以是全 URL(如'http://www.3testing.com/')或者是相对 URL(如'/login/')。返回 302 状态码。

(2) HttpResponsePermanentRedirect:同 HttpResponseRedirect 一样,但是返回的是永久重定向。返回 301 状态码。

(3) HttpResponseNotModified:构造函数不需要参数。使用此命令指定页面自用户上次请求后未被修改。返回 304 状态码。

(4) HttpResponseBadRequest:返回 400 状态码。

(5) HttpResponseNotFound:返回 404 状态码。

(6) HttpResponseForbidden:返回 403 状态码。

(7) HttpResponseNotAllowed:返回 405 状态码。它需要一个必需的参数,一个允许方法的 List(如[' GET','POST'])。

(8) HttpResponseGone:返回 410 状态码。

(9) HttpResponseServerError:返回 500 状态码。

2.5　setting.py 的配置

2.5.1　中间件介绍

前面介绍了中间件 django.middleware.csrf.CsrfViewMiddleware，下面介绍其他几个中间件。

（1）认证支持中间件：django.contrib.auth.middleware.AuthenticationMiddleware。

这个中间件激活认证支持功能，它在每个传入的 HttpRequest 对象中添加了代表当前登录用户 Request.user 的属性，即 Request.UserHostAddress 属性与 Request.UserLanguages 属性。Request.UserHostAddress 属性可以获得访问者的 IP 地址，而 Request.UserLanguages 属性可以获得访问者浏览器支持的语言，通过这个属性就可以实现不同语言的人显示不同语言的页面。

（2）通用中间件：django.middleware.common.CommonMiddleware。

这个中间件为完美主义者提供了一些便利：禁止 DISALLOWED_USER_AGENTS 列表中设置的 user agent 访问。一旦提供，这一列表应当由已编译的正则表达式对象组成，这些对象用于匹配传入的 request 请求头中的 user-agent 域。下面这个例子来自某个配置文件片段。

```
import re
DISALLOWED_USER_AGENTS = (
    re.compile(r'^OmniExplorer_Bot'),
    re.compile(r'^Googlebot')
)
```

请注意 import re，因为 DISALLOWED_USER_AGENTS 要求其值为已编译的正则表达式（也就是 re.compile() 的返回值）。配置文件是常规的 Python 文件，所以其中包括 Python import 语句不会有任何问题。

依据 APPEND_SLASH 和 PREPEND_WWW 的设置执行 URL 重写：如果 APPEND_SLASH 为 True，那些尾部没有斜杠的 URL 将被重定向到添加了斜杠的相应 URL，除非 path 的最末组成部分包含点号。因此，foo.com/bar 会被重定向到 foo.com/bar/，但是 foo.com/bar/file.txt 将以不变形式通过。

如果 PREPEND_WWW 为 True，那些缺少先导 www. 的 URLs 将会被重定向到含有先导 www. 的相应 URL 上。

这两个选项都是为了规范化 URL。其后的哲学是每个 URL 都应当且只应当存在于一处。从技术上来说，URL：example.com/bar 与 example.com/bar/、www.example.com/bar/都互不相同。搜索引擎编目程序把它们视为不同的 URL，这将不利于该站点的搜索引擎排名，因此这里的最佳实践是将 URL 规范化。

依据 USE_ETAGS 的设置处理 Etag：ETags 是 HTTP 级别上按条件缓存页面的优化机制。如果 USE_ETAGS 为 True，Django 针对每个请求以 MD5 算法处理页面内容，从而得到 Etag，在此基础上，Django 将在适当情形下处理并返回 Not Modified 回应（或者设置

response 头中的 Etag 域)。

（3）压缩中间件：django.middleware.gzip.GZipMiddleware。

这个中间件可以自动处理 gzip 压缩文件的浏览器（包括所有的浏览器）自动压缩返回的内容，这样将大大减少 Web 服务器所耗用的带宽。但是，代价是压缩页面需要一些额外的处理时间。

（4）条件化的 GET 中间件：django.middleware.http.ConditionalGetMiddleware。

这个中间件对条件化 GET 操作提供支持。如果 Response 头中包括 Last-Modified 或 ETag 域，并且 Request 头中包含 If-None-Match 或 If-Modified-Since 域，且两者一致，则该 Response 将被 response 304（没有改变）取代。对 ETag 的支持依赖于 USE_ETAGS 配置及事先在 Response 头中设置的 ETag 域。

此外，这个中间件也将删除处理 HEAD request 时所生成 Response 中的任何内容，并在所有 Request 的 Response 头中设置 Date 和 Content-Length 域。

（5）反向代理支持[①]（X-Forwarded-For 中间件）：django.middleware.http.SetRemoteAddrFromForwardedFor。

在 Request.META['HTTP_X_FORWARDED_FOR']存在的前提下，它根据其值来设置 request.META['REMOTE_ADDR']。在站点位于某个反向代理之后，每个 Request 的 REMOTE_ADDR 都被指向 127.0.0.1。

（6）会话支持中间件：django.contrib.sessions.middleware.SessionMiddleware。

这个中间件激活会话支持功能，它们互相配合，以缓存每个基于 Django 的页面。

（7）事务处理中间件：django.middleware.transaction.TransactionMiddleware。

这个中间件把数据库的 COMMIT 或 ROLLBACK 绑定到 request/response 处理阶段。如果 view 函数成功执行，则发出 COMMIT 指令。如果 view 函数抛出异常，则发出 ROLLBACK 指令。这样，程序员就不用在程序中专门处理 COMMIT 或 ROLLBACK 了。

2.5.2 其他配置介绍

setting.py 的其他配置见表 2-1。该表格来源于 Django 的官方文档，略作改动。

表 2-1 setting.py 的其他配置

设 置	默认值	介 绍	案 例

ABSOLUTE_URL_OVERRIDES	{}(空字典)	一个字典映射 "app_label.module_name" 字符串到一个函数，该函数接受一个 model 对象作为参数并返回它的 URL。这是在安装上覆盖 get_absolute_url() 方法的一种方式	ABSOLUTE_URL_OVERRIDES ={
			'blogs.blogs': lambda o:
			"/blogs/%s/" % o.slug,
			'news.stories': lambda o:
			"/stories/%s/%s/"%
			(o.pub_year, o.slug),
			}

[①] 反向代理(Reverse Proxy)方式是指以代理服务器来接受 Internet 上的连接请求，然后将请求转发给内网络上的服务器，并将从服务器上得到的结果返回给 Internet 上请求连接的客户端，此时代理服务器对外就表现为一个反向代理服务器。

续表

设　　置	默认值	介　　绍	案　　例
ADMIN_FOR	()(空的元祖)	用于 admin-site settings 模块，如果当前站点是 admin,则它是一个由 settings 模块组成的元组（类似'foo.testing.test' 这样的格式） admin 站点在 models、views 及 template tags 的自动内部文档中使用该设置	
ADMIN_MEDIA_PREFIX	'/media/'	管理介质的 URL 前缀——CSS、JavaScript 和图像,确保使用斜线	
ADMINS	()(空的元祖)	ADMINS 是一个二元元组,记录开发人员的姓名和 E-mail,当 DEBUG 为 False,而 views 发生异常的时候,发 E-mail 通知这些开发人员。该元组的每个成员的格式为(Full name, Email address)	(('John','john@example.com'), ('Mary','mary@example.com'))
ALLOWED_HOST	[](空的列表)	为了限定请求中的 host 值,以防止黑客构造包来发送请求。只有在列表中的 host 才能访问。强烈建议不要使用 * 通配符去配置。另外,当 DEBUG 设置为 False 的时候,必须配置这个配置,否则就会抛出异常	ALLOWED_HOSTS = [　'.example.com', # Allow domain and subdomains 　'.example.com.', # Also allow FQDN and subdomains]
ALLOWED_INCLUDE_ROOTS	()(空的元祖)	一个字符串元组,只有以列表中的元素为前缀的模板,Django 才可以以"{% ssi %}"形式访问。出于安全考虑,在不应该访问时,即使是模板的作者,也不能访问这些文件	如果 ALLOWED_INCLUDE_ROOTS 是 ('/home/html ', '/var/www'),那么 {% ssi /home/html/foo.txt %} 可以正常工作,不过 {% ssi /etc/passwd %} 却不能
APPEND_SLASH	True	是否给 URL 添加一个结尾的斜线。只有安装了 CommonMiddleware 之后,该选项才起作用,参阅 PREPEND_WWW	

续表

设置	默认值	介绍	案例

AUTH_PASSWORD_VALIDATORS		参见 https://docs.djangoproject.com/en/1.11/ref/settings/#auth-password-validators	[
			  {
			    'NAME': 'django.contrib.auth.password_validation.UserAttributeSimilarityValidator',
			  },
			  {
			    'NAME': 'django.contrib.auth.password_validation.MinimumLengthValidator',
			  },
			  {
			    'NAME': 'django.contrib.auth.password_validation.CommonPasswordValidator',
			  },
			  {
			    'NAME': 'django.contrib.auth.password_validation.NumericPasswordValidator',
			  },
			]

CACHE_BACKEND	'simple://'	后端使用的 cache，参阅 cache docs	
CACHE_MIDDLEWARE_KEY_PREFIX	''(空的字符串)	cache 中间件使用的 cache key 前缀，参阅 cache docs	
DATABASE_ENGINE	'postgresql'	后端使用的数据库引擎：'postgresql'、'mysql'、'sqlite3' 或 'ado_mssql' 中的任意一个	
DATABASE_HOST	''(空的字符串)	数据库所在的主机。空的字符串意味着 localhost。SQLite 不需要该项。如果使用 MySQL 并且该选项的值以一个斜线('/') 开始，MySQL 则通过一个 UNIX socket 连接到指定的 socket	DATABASE_HOST = '/var/run/mysql' 如果使用 MySQL 并且该选项的值不是以斜线开始，那么该选项的值就是主机的名字
DATABASE_NAME	''(空的字符串)	要使用的数据库名字。对 SQLite，它必须是一个数据库文件的全路径名字	
DATABASE_PASSWORD	''(空的字符串)	连接数据库需要的密码。SQLite 不需要该项	
DATABASE_PORT	''(空的字符串)	连接数据库所需的数据库端口，空的字符串表示默认端口。SQLite 不需要该项	

续表

设置	默认值	介绍	案例
DATABASE_USER	' '(空的字符串)	连接数据库时所需要的用户名。SQLite 不需要该项	
DATE_FORMAT	'N j, Y'（如 Feb. 21, 2017）	在 Django admin change-list 页对日期字段使用的默认日期格式，系统中的其他部分也可能使用该格式，参阅 allowed date format strings	
DEBUG	False	DEBUG 配置为 True 的时候，会暴露出一些出错信息或者配置信息，以方便调试。注意，在上线的时候应该将其关掉，防止配置信息或者敏感出错信息泄露	
DEFAULT_CHARSET	'utf-8'	如果一个 MIME 类型没有特别指定，对所有 HttpResponse 对象将应用该默认字符集。使用 DEFAULT_CONTENT_TYPE 来构建 Content-Type 头	
DEFAULT_CONTENT_TYPE	'text/html'	如果一个 MIME 类型没有特别指定，对所有 HttpResponse 对象将应用该默认 content type。使用 DEFAULT_CHARSET 来构建 Content-Type 头	
DEFAULT_FROM_EMAIL	'webmaster@localhost'	用于发送（站点自动生成的）管理邮件的默认 E-mail 邮箱	
DISALLOWED_USER_AGENTS	()（空的元组）	一个编译的正则表达式对象列表，用于表示一些用户代理的字符串。这些用户代理将被禁止访问系统中的任何页面。使用这个页面的有机器人或网络爬虫。只有安装 CommonMiddleware 后，这个选项才有用	
EMAIL_HOST	'localhost'	用来发送 E-mail 的主机	
EMAIL_HOST_PASSWORD	' '(空的字符串)	EMAIL_HOST 中定义的 SMTP 服务器使用的密码。如果为空，Django 是不会进行认证的	
EMAIL_HOST_USER	' '(空的字符串)	EMAIL_HOST 中定义的 SMTP 服务器使用的用户名。如果用户名为空，Django 是不会进行认证的	
EMAIL_PORT	25	EMAIL_HOST 中指定的 SMTP 服务器所使用的端口号	

续表

设 置	默认值	介 绍	案 例
EMAIL_SUBJECT_PREFIX	'[Django]'	django.core.mail.mail_admins 或 django.core.mail.mail_managers 发送邮件的主题前缀	
ENABLE_PSYCO	False	如果允许 Psyco,将使用 Pscyo 优化 Python 代码。需要 Psyco 模块	
IGNORABLE_404_ENDS	('mail.pl', 'mailform.pl', 'mail.cgi', 'mailform.cgi', 'favicon.ico', '.php')	参阅 IGNORABLE_404_STARTS	
IGNORABLE_404_STARTS	('/cgi-bin/', '/_vti_bin/', '/_vti_inf')	一个字符串元组。以该元组中元素为开头的 URL 必须被 404 Emailer 忽略,参阅 SEND_BROKEN_LINK_EMAILS 和 IGNORABLE_404_ENDS	
INSTALLED_APPS	()(空的元组)	一个字符串元组,内容是本 Django 安装中的所有应用。每个字符串应该是一个包含 Django 应用程序 Python 包的路径全称,django-admin.py startapp 会自动往其中添加内容	INSTALLED_APPS = ['django.contrib.admin', 'django.contrib.auth', 'django.contrib.contenttypes', 'django.contrib.sessions', 'django.contrib.messages', 'django.contrib.staticfiles', 'goods']
INTERNAL_IPS	()(空的元组)	一个 IP 地址的元组(字符串形式),当 DEBUG 为 True 时,可用于设置允许访问 debug toolbar 的 IP 地址	
JING_PATH	'/usr/bin/jing'	"Jing"执行文件路径全名。Jing 是一个 RELAXNG 校验器,Django 使用它对 model 的 XMLField 进行验证,参阅 http://www.thaiopensource.com/relaxng/jing.html	
LANGUAGE_CODE	'en-us'	表示默认语言的一个字符串,必须是标准语言格式	U.S. English 就是 "en-us"。如果用汉语,则必须设置为 "zh-cn"

续表

设　　置	默认值	介　　绍	案　　例
LANGUAGES	一个元组（内容为所有可用语言）	一个二元元组＜格式为（语言代码，语言名称）＞的元组。该设置用于选择可用语言	LANGUAGES ＝ （ ('bn', _('Bengali'))， ('cs', _('Czech'))， ('cy', _('Welsh'))， ('da', _('Danish'))， ('de', _('German'))， ('en', _('English'))， ('es', _('Spanish'))， ('fr', _('French'))， ('gl', _('Galician'))， ('is', _('Icelandic'))， ('it', _('Italian'))， ('no', _('Norwegian'))， ('pt-br', _('Brazilian'))， ('ro', _('Romanian'))， ('ru', _('Russian'))， ('sk', _('Slovak'))， ('sr', _('Serbian'))， ('sv', _('Swedish'))， ('zh-cn', _('Simplified Chinese'))，)
MANAGERS	ADMINS	和 ADMINS 类似，并且结构一样，当出现'broken link'的时候给 manager 发邮件	
MEDIA_ROOT	' '(空的字符串)	一个绝对路径，用于保存媒体文件	"/home/media/media.lawrence.com/"
MEDIA_URL	' '(空的字符串)	处理媒体服务的 URL（媒体文件来自 MEDIA_ROOT）	"http://media.lawrence.com"
MIDDLEWARE_CLASSES	见本行例子	Web 应用中需要加载的一些中间件列表，是一个一元数组。里面是 Django 自带的或者定制中间件的包路径	['django.middleware.security.SecurityMiddleware', 'django.contrib.sessions.middleware.SessionMiddleware', 'django.middleware.common.CommonMiddleware', 'django.middleware.csrf.CsrfViewMiddleware', 'django.contrib.auth.middleware.AuthenticationMiddleware', 'django.contrib.messages.middleware.MessageMiddleware', 'django.middleware.clickjacking.XFrameOptionsMiddleware',]

续表

设　置	默认值	介　绍	案　例
PASSWORD_HASHER	见本行例子	这个配置是在使用 Django 自带的密码加密函数的时候会使用的加密算法的列表。默认为本行例子。在这个例子中使用 PBKDF2① 加密算法。所以,在使用 make_password, check_password, is_password_unable 等密码加解密函数的时候,需要在 setting.py 文件中添加这个 list,推荐使用默认配置的算法	PASSWORD_HASHERS = ('django.contrib.auth.hashers.PBKDF2PasswordHasher', 'django.contrib.auth.hashers.PBKDF2SHA1PasswordHasher', 'django.contrib.auth.hashers.BCryptSHA256PasswordHasher', 'django.contrib.auth.hashers.BCryptPasswordHasher', 'django.contrib.auth.hashers.SHA1PasswordHasher', 'django.contrib.auth.hashers.MD5PasswordHasher', 'django.contrib.auth.hashers.CryptPasswordHasher',)
PREPEND_WWW	False	是否为没有"www."前缀的域名添加"www."前缀。当且仅当安装有 CommonMiddleware 后,该选项才有效,参阅 APPEND_SLASH	
ROOT_URLCONF	Not defined	一个字符串,表示根 URLconf 的模块名	'ebusiness.urls'
SECRET_KEY	' '(空的字符串)	一个密码,用于为密码哈希算法提供一个种子。将其设置为一个随机字符串,并且越长越好。django-admin.py startproject 会自动创建一个密码字符串	
SEND_BROKEN_LINK_EMAILS	False	当有人从一个有效 Django-powered 页面访问另一个 Django-powered 页面时发现 404 错误(也就是发现一个死链接时),是否发送一封邮件给 MANAGERS。当且仅当安装有 CommonMiddleware 时,该选项才有效,参阅 IGNORABLE_404_STARTS 和 IGNORABLE_404_ENDS	
SERVER_EMAIL	'root@localhost'	用来发送错误信息的邮件地址,如发送给 ADMINS 和 MANAGERS 的邮件	

---

① PBKDF2 应用一个伪随机函数,以导出密钥。导出密钥的长度本质上是没有限制的(但是,导出密钥的最大有效搜索空间受限于基本伪随机函数的结构)。

续表

设 置	默认值	介 绍	案 例
SESSION_COOKIE_AGE	1209600（2 周，以秒计）	session cookies 的生命周期，以秒计	
SESSION_COOKIE_DOMAIN	None	session cookies 有效的域。将其值设置为类似 ".3testing.com" 这样的 cookie，就可以跨域生效，或者使用 None 作为一个标准的域 cookie	
SESSION_COOKIE_NAME	'sessionid'	session 使用的 cookie 名字	
SESSION_SAVE_EVERY_REQUEST	False	是否每次请求都保存 session	
SITE_ID	Not defined	是一个整数，表示 django_site 表中的当前站点。当一个数据包含多个站点数据时，程序可以据此 ID 访问特定站点的数据	
TEMPLATE_CONTEXT_PROCESSORS	见本行案例	一个元组，如果当前模块下的所有视图都需要共同变量，就想到了利用 context_processors。例如，每次返回 response，都要加一样的变量，如 {'user': username, 'role': role}，这时采用 context_processors 可以在每次返回时不用带 {'user': username, 'role': role}，而是将这些变量写到 context_processors 里面	("django.core.context_processors.auth", "django.core.context_processors.debug", "django.core.context_processors.i18n")
TEMPLATE_DEBUG	False	一个布尔值，用于开关模板调试模式。该值设置为 True 时，如果有任何 TemplateSyntaxError，一个详细的错误报告信息页将被显示。这个报告包括有关的模板片段，相应的行会自动高亮。 注意，Django 仅在 DEBUG 为 True 时显示这个信息页面	
TEMPLATE_DIRS	()（空的元组）	模板源文件目录列表，按搜索顺序排列。注意，要使用 UNIX 风格的前置斜线（即'/'），即使在 Windows 上也一样	
TEMPLATE_LOADERS	见本行案例	一个元素为可调用对象（字符串形式）的元组。这些对象知道如何从各种源中导入 templates	('django.template.loaders.filesystem.load_template_source',)

续表

设　置	默认值	介　绍	案　例
TEMPLATE_STRING_IF_INVALID	' '(空的字符串)	输出文本,作为一个字符串。模板系统将会在出错(如拼写错误)时使用该变量	
TIME_FORMAT	'P'	Django admin change-list 使用默认时间格式。有可能系统的其他部分也使用该格式,参阅 DATE _ FORMAT 和 DATETIME_FORMAT	
TIME_ZONE	'America/Chicago'	一个表示当前时区的字符串。Django 在这里设置转换所有的日期/时间——并不考虑服务器的时区设置。例如,一台服务器可以服务多个 Django-powered 站点,每个站点使用一个独立的时区设置	'Asia/Shanghai PRC'
USE_ETAGS	False	一个布尔值,指定是否输出"Etag"头。这个选项可以节省网络带宽,但损失性能。只有安装了 CommonMiddleware 后,这个选项才会起作用	

2.5.3　自定义静态文件

一些图片文件、CSS 文件、JavaScript 文件可以放在一个专门的地方,下面介绍如何进行设置。首先打开 setting.py,然后进行如下设置。

```
...
STATIC_URL = '/static/'
BASE_DIR = os.path.dirname(os.path.dirname(os.path.abspath(__file__)))

STATICFILES_FINDERS = (
    "django.contrib.staticfiles.finders.FileSystemFinder",
    "django.contrib.staticfiles.finders.AppDirectoriesFinder"
)

STATICFILES_DIRS = (
    os.path.join(BASE_DIR,"static"),
)
...
```

(1) STATIC_URL = '/static/'规定了静态文件的 URL。

(2) STATICFILES_FINDERS 规定了静态文件的查找顺序和内容,它首先通过 django. contrib. staticfiles. finders. FileSystemFinder 读取是否存在规定的 STATICFILES_

DIRS,如果不存在,则通过 django.contrib.staticfiles.finders.AppDirectoriesFinder 在每个应用中查找有没有静态文件目录 static。

(3) STATICFILES_DIRS 中的 os.path.join(BASE_DIR,"static")表示静态路径的目录在 BASE_DIR \ static 下。在模板文件中调用的时候,头部需要加上{%load staticfiles%}。

具体使用的时候用类似于这样的方法:<link href="{%static 'css/signin.css'%}" rel="stylesheet">首先由{%%}括起来,然后在{%后面跟上 static,后面是一个字符串,字符串中是具体 static 后的路径+文件名(一般地,static 目录下会生成四个目录:js 存放 javascript 文件、image 存放图片文件、css 存放 css 文件、admin 存放 admin 管理员模块所用的文件。admin 目录下又分四个目录:js 存放 javascript 文件、img 存放图片文件(注意,这里不是 image 目录名称)、css 存放 css 文件、fonts 存放字体文件)。

接下来在 url.py 中加入下列语句。

```
import os

BASE_DIR =os.path.dirname(os.path.dirname(os.path.abspath(__file__)))
...
url(r'^static/(?P<path>.*)',static.serve,{'document_root':os.path.join(BASE_DIR,'static')}),
...
```

这里,查看文件登录页面 index.html 的模板文件。

```
{%load staticfiles%}
...
    <!--Bootstrap core CSS -->
    <link href="{%static 'css/signin.css'%}" rel="stylesheet">
    <!--Custom styles for this template -->
    <link href="{%static 'css/bootstrap.min.css'%}" rel="stylesheet">
    <link href="{%static 'css/my.css'%}" rel="stylesheet">
...
```

2.5.4 案例

下面给出本书项目使用后 setting.py 的设置。

```
"""
Django settings for ebusiness project.

Generated by 'django-admin startproject' using Django 1.11.4.

For more information on this file, see
https://docs.djangoproject.com/en/1.11/topics/settings/
```

```
For the full list of settings and their values, see
https://docs.djangoproject.com/en/1.11/ref/settings/
"""

import os

#Build paths inside the project like this: os.path.join(BASE_DIR, ...)
BASE_DIR =os.path.dirname(os.path.dirname(os.path.abspath(__file__)))


#Quick-start development settings -unsuitable for production
#See https://docs.djangoproject.com/en/1.11/howto/deployment/checklist/

#SECURITY WARNING: keep the secret key used in production secret!
SECRET_KEY ='ok4*wia=(o6ycxkmhfls6% wydyl@yg-i$u-s=$b@y9#wjzxlrk'

#SECURITY WARNING: don't run with debug turned on in production!
DEBUG =False

ALLOWED_HOSTS =[]


#Application definition

INSTALLED_APPS =[
    'django.contrib.admin',
    'django.contrib.auth',
    'django.contrib.contenttypes',
    'django.contrib.sessions',
    'django.contrib.messages',
    'django.contrib.staticfiles',
    'goods'
]

MIDDLEWARE =[
    'django.middleware.security.SecurityMiddleware',
    'django.contrib.sessions.middleware.SessionMiddleware',
    'django.middleware.common.CommonMiddleware',
    'django.middleware.csrf.CsrfViewMiddleware',
    'django.contrib.auth.middleware.AuthenticationMiddleware',
    'django.contrib.messages.middleware.MessageMiddleware',
    'django.middleware.clickjacking.XFrameOptionsMiddleware',
]
```

```python
ROOT_URLCONF = 'ebusiness.urls'

TEMPLATES = [
    {
        'BACKEND': 'django.template.backends.django.DjangoTemplates',
        'DIRS': [],
        'APP_DIRS': True,
        'OPTIONS': {
            'context_processors': [
                'django.template.context_processors.debug',
                'django.template.context_processors.request',
                'django.contrib.auth.context_processors.auth',
                'django.contrib.messages.context_processors.messages',
            ],
        },
    },
]

WSGI_APPLICATION = 'ebusiness.wsgi.application'


# Database
# https://docs.djangoproject.com/en/1.11/ref/settings/#databases

DATABASES = {
    'default': {
        'ENGINE': 'django.db.backends.sqlite3',
        'NAME': os.path.join(BASE_DIR, 'db.sqlite3'),
    }
}


# Password validation
# https://docs.djangoproject.com/en/1.11/ref/settings/#auth-password-validators


AUTH_PASSWORD_VALIDATORS = [
    {
        'NAME': 'django.contrib.auth.password_
        validation.UserAttributeSimilarityValidator',
    },
    {
        'NAME': 'django.contrib.auth.password_validation.MinimumLengthValidator',
    },
```

```
    {
        'NAME': 'django.contrib.auth.password_validation.CommonPasswordValidator',
    },
    {
        'NAME': 'django.contrib.auth.password_validation.NumericPasswordValidator',
    },
]


# Internationalization
# https://docs.djangoproject.com/en/1.11/topics/i18n/

LANGUAGE_CODE = 'en-us'

TIME_ZONE = 'UTC'

USE_I18N = True

USE_L10N = True

USE_TZ = True


# Static files (CSS, JavaScript, Images)
# https://docs.djangoproject.com/en/1.11/howto/static-files/

STATIC_URL = '/static/'

STATICFILES_FINDERS = (
    "django.contrib.staticfiles.finders.FileSystemFinder",
    "django.contrib.staticfiles.finders.AppDirectoriesFinder"
)

STATICFILES_DIRS = (
    os.path.join(BASE_DIR,"static"),
)

ALLOWED_HOSTS=" * "
```

2.6 session 和 cookie

根据 1.3.5 节的介绍，已经了解了 session 和 cookie 是为了解决 HTTP 自身的无状态性。这里主要介绍 Django 是如何实现 session 和 cookie 的。

2.6.1 session

Django 的 session 服务器端是存储在数据库表中的,Django 默认的数据库是 SQLite3。在 Django 中要使用 session,必须先建立数据库。

首先检查目录 ebusiness 下的 settings.py 文件(这个文件默认是 SQLite3 的配置文件)。

```
...
#Database
#https://docs.djangoproject.com/en/1.11/ref/settings/#databases

DATABASES = {
    'default': {
        'ENGINE': 'django.db.backends.sqlite3',
        'NAME': os.path.join(BASE_DIR, 'db.sqlite3'),
    }
}...
```

然后在上一级目录下通过命令/ebusiness>python manage.py migrate 创建数据库。运行结果如图 2-4 所示。

图 2-4 运行结果

运行完毕后,在当前目录下生成一个名为 db.sqlite3 的数据库文件,登录系统,用第三方工具(如 SQLite manage)打开这个数据库,可以看到如图 2-5 所示的数据。

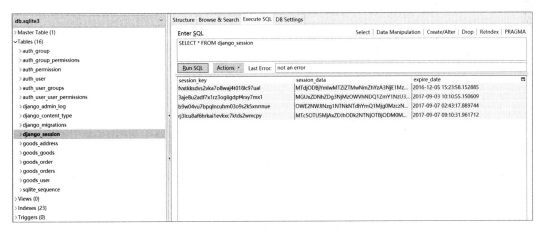

图 2-5 数据库中的 session 数据

1. 建立 session

```
request.session('key') = value
```

这里,key 为 session 的键,value 为 session 的值。例如,key 为 username,value 为 Jerry,即 request.session('username') = Jerry,表示 session 中 username 的值为 Jerry。下面讨论如何获取 session。

2. 获取 session

```
var = request.session.get(key,'')
```

其中,var 为获取到的 session 的 value。提到 session,读者肯定会想起系统的登录,登录以后,系统会把登录信息作为一个 session 进行保存,它要比 cookie 安全得多。下面来看 Django 程序是如何完成登录功能的。

```python
...
from django.http import HttpResponse,HttpResponseRedirect
...
def login_action(request):
    if request.method == "POST":
        uf = LoginForm(request.POST)
        if uf.is_valid():
            #寻找名为 "username"和"password"的 POST 参数,如果参数没有提交,
            #则返回一个空的字符串
            username = uf.cleaned_data['username']
            password = uf.cleaned_data['password']
            #判断输入的数据是否为空
            if username == '' or password == '':
                return render(request,"index.html",{'uf':uf,"error":"用户名和密码不能为空"})
            else:
                #判断用户名和密码是否准确
                ...
                if user:
                    response = HttpResponseRedirect('/goods_view/')
                    #登录成功,跳转查看商品信息
                    request.session['username'] = username
                    #将 session 信息写到服务器
                    return response
                else:
                    return render(request,"index.html",{'uf':uf,"error":"用户名或者密码错误"})
    else:
        uf = LoginForm()
    return render_to_response('index.html',{'uf':uf})
...
```

语句 request.session['username'] = username 正是把用户名作为 session 的 username 值进行存储,然后在需要登录的页面中加入如下代码。

```
...
username = str(request.session.get('username',''))
if username is not None:
...
```

如果 username 在 session 中被存储,则程序会返回,否则返回 NULL。为了更安全,从 session 中获取 username 后,程序再到数据库中进行查询,确保 session 存储的是注册过的用户,只允许这样的用户访问需要登录的页面。

注销登录(也就是登出操作)可以用如下程序来实现。

```
...
#用户登出
def logout(request):
    response = HttpResponseRedirect('/index/') #登出成功后跳转到登录页面
    request.session['username'] = "" #将 username 的 session 信息清空,然后写入服务器
    return response
...
```

2.6.2 cookie

1. 建立 cookie

```
...
response = HttpResponseRedirect('/url/')
response.set_COOKIE(key,value,time)
return response
...
```

(1) response = HttpResponseRedirect('/url/'):在相应的 URL 中设置 cookie。

(2) response.set_COOKIE(key,value,time):设置 cookie,key 为 cookie 的键;value 为 cookie 的值;time 为 cookie 的存活期限,单位为秒。

所以,建立 cookie 的语句为

```
response.set_COOKIE(key,value,time)
```

2. 获得 cookie

```
...
var=request.COOKIE.get('key', '')
...
```

其中,key 为 cookie 的名字,var 为 cookie 的值。因此,获取 cookie 的语句为

```
var = request.COOKIE.get('key', '')
```

3. 修改 cookie

修改 cookie 就是重新设置 key 所对应的值,方法如下。

```
...
response = HttpResponseRedirect('/url/')
response.set_COOKIE(key,new_value,time)
return response
...
```

4. 删除 cookie

把 cookie 的时间设置为小于或者等于 0。

```
...
response = HttpResponseRedirect('/url/')
response.set_COOKIE(key,new_value,0)
return response
...
```

2.6.1 节提到用户登录,如果登录一个网站,会随机产生一个名为 sessionid 的 cookie,也可以通过判断是否存在这个 sessionid 来避免没有登录的非法用户进入系统,具体代码如下。

```
...
        username = request.session.get('username','')
        #获得所有的 cookie
        cookie_list = request.COOKIES
        if ("sessionid" in cookie_list) and (username is not None):
            return username
        else:
            return ""
...
```

其中,request.COOKIES 为获得所有的 cookie。在 goods 目录下建立一个名为 util.py 的文件,将这个验证程序封装起来。

```
...
class Util():
    def check_user(self,request):
        #从 cookies 中取出 username
        username = request.session.get('username','')
        #获得所有的 cookie
        cookie_list = request.COOKIES
        if ("sessionid" in cookie_list) and (username is not None):
            return username
        else:
            return ""
...
```

然后在 views.py 中这样调用:

```
...
    util =Util()
    username =util.check_user(request)
    if username=="":
        uf =LoginForm()
        return render(request,"index.html",{'uf':uf,"error":"请登录后再进入"})
    else:
        #做你正常的工作
...
```

最后总结一下 cookie、session 和 sessionid 的区别与联系。cookie 的内容主要包括：名字、值、过期时间、路径和域（路径和域也可以通过方法 set_COOKIE() 设置，由于它们不经常使用，这里不进行介绍）。路径和域一起构成 cookie 的作用范围。若不设置过期时间，则表示这个 cookie 的生命周期为浏览器会话期间，关闭浏览器窗口，cookie 就消失。这种生命周期为浏览器会话期的 cookie 被称为会话 cookie。会话 cookie 一般不存储在硬盘上，而是保存在内存里，当然，这种行为并不是规范规定的。若设置了过期时间，浏览器就会把 cookie 保存到硬盘上，关闭后再次打开浏览器，这些 cookie 仍然有效，直到超过设定的过期时间。存储在硬盘上的 cookie 可以在不同的浏览器进程间共享，例如两个浏览器窗口。而对于保存在内存里的 cookie，不同的浏览器有不同的处理方式。

session 可以理解为一个 cookie，它的名字（name）是存在浏览器端的，但是它的值（value）是存在服务器端的，这就是前面提到的使用 session 比使用 cookie 登录更安全的原因。

当程序需要为某个客户端的请求创建一个 session 时，服务器首先要检查这个客户端的请求里是否已包含了一个名为 session 的标识——称为 sessionid，如果已包含，则说明以前已经为此客户端创建过 session，服务器就按照 sessionid 把这个 session 检索出来使用（当然，一旦检索不到，就会新建一个 session），如果客户端请求不包含 sessionid，就为此客户端创建一个 session 并且生成一个与此 session 相关联的 sessionid，sessionid 的值应该是一个既不会重复，又不容易被找到的随机字符串，这个 sessionid 在本次响应中将被返回给客户端保存。

2.6.3 Django 的用户登录和注册机制

Django 系统本身提供登录机制，但是，遗憾的是它没有提供注册机制，所以只有系统不需要实现注册功能，如用户签到系统、带密码的个人便签系统，可以让注册操作由特定的一到多个管理员，甚至是用户本人来进行操作，这样就可以使用 Django 自己提供的更加安全的登录机制了[1]（本书仅对关于 Django 自身的登录和注册机制知识进行简单介绍，如果读者对此非常感兴趣，建议查看参考文献[5]中的第 2 章）。本书采用自定义的方法来实现用户的登录和注册。

首先通过命令\ebusiness＞ python manage.py createsuperuser 创建一个超级用户，如

---

① Django 没有提供注册机制让许多人感到非常的遗憾，所以出现了一些第三方的系统，比较著名的是 django-registration。大家可以在网站查找这个资料，本书不进行介绍。

图 2-6 所示。

图 2-6　创建一个超级用户

这里需要注意的是密码不能太简单（如全是数字）并且长度不能少于 8 位。然后在浏览器中输入 http://127.0.0.1:8000/admin/用刚建立的用户登录即可看见如图 2-7 所示的登录后主页面。

图 2-7　登录后主页面

下面来看如何通过代码实现用户的登录。

```
...
username = uf.cleaned_data['username']
password = uf.cleaned_data['password']
user = auth.authenticate(username=username, password=password)
if user is not None:
    auth.login(request, user)
    request.session('user') = username
    response = HttpResponseRedirect('/url/')
    return response
...
```

对于需要进行登录才可以看到的网站，只需在方法前面加上 @login_required 就可以了，例如

```
...
@login_required
def good_view(request):
    ...
    ...
```

但是，当访问 http://127.0.0.1:8000/good_view/的时候，系统会报 404 错误，只要在 ebusiness 目录下的 url.py 文件中加入如下代码即可。

```
...
urlpatterns = [
    url(r'^index/$ ', views.index),
    url(r'^$ ', views.index),
    url(r'^account/login/$ ', views.index)
    ...
...
```

这时只要输入下列三行中的任一行代码，
(1) http://127.0.0.1:8000/。
(2) http://127.0.0.1:8000/good_view/。
(3) http://127.0.0.1:8000/index/。

都会显示登录界面。在这里需要建立 account/login URL 的原因是因为当输入 http://127.0.0.1:8000/good_view/时，由于没有登录，系统会转向一个默认名为 account/login 的 URL，如果在 url.py 中没有定义，系统就会显示 404 错误。

既然提到登录，就要讨论一下登出。登出的代码也很简单，如下所示。

```
...
@login_required
def logout(request)
    auth.logout(request)
    response = HttpResponseRedirect('/index/')
    return response
...
```

2.7 Django 的 MTV 开发模式框架

开发动态网站，大家都会想起 MVC 模式。MVC 模式是由程序员 Trygve Reenskaug 于 1978 年提出的，是施乐帕罗奥多研究中心（Xerox PARC）在 20 世纪 80 年代为程序 Smalltalk 发明的一个软件开发框架。Django 是通过 MTV 的框架进行开发的。所谓 MTV，即 Model、Template 和 View 三个英文字母的缩写组合，是一个 Django 自定义的 MVC 框架体系。

(1) Model：模型，是建立程序与数据库之间的纽带，主要通过应用（如 goods）目录下的 models.py 实现。

(2) Template：模板，如前面的介绍，在应用（如 goods）目录下建立一个 templates 目录，在这个目录中存放应用需要的 HTML 文档模板。

(3) View：视图，是业务逻辑层，是连接 Model 和 Template 的纽带，主要通过应用（如 goods）目录下的 views.py 来实现。

这里，MTV 开发框架中的 M、T、V 分别对应 MVC 开发框架中的 M、V、C，如图 2-8 所示。

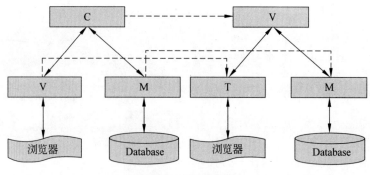

图 2-8　MTV 与 MVC

2.8　Django 的模型与数据库的管理

2.8.1　Django 的数据库

下面介绍 Django 的模型与数据库的管理，正如前面提到 Django 默认的数据库是 SQLite3，但是它也可以支持其他三种数据库，分别是 PostgreSQL、MySQL 和 Oracle。

(1) SQLite3：建议使用 Python 2.5 以上，否则可能出现兼容性问题。

(2) PostgreSQL：下载 psycopg 开发包，建议使用 psycopg 2。

(3) MySQL：MySQL 4.0 或更高的版本。

(4) Oracle：Oracle 9i 或更高版本＋cx_Oracle 库(4.3.1 版本或更高版本，但是不要使用 5.0 版本)。

关于 SQLite3 在 setting.py 中的配置，本书在 2.6.1 节中已进行了介绍，下面系统介绍所有数据库的配置方法。

```
...
#Database
#https://docs.djangoproject.com/en/1.11/ref/settings/#databases

DATABASES = {
    'default': {
        'ENGINE': 'django.db.backends.mysql',
        'NAME': 'mydatabase',
        'USER': 'mydatabaseuser',
        'PASSWORD': 'mypassword',
        'HOST': '127.0.0.1',
        'PORT': '3306',
    }
}
...
```

下面介绍参数。

(1) ENGINE：指定数据库驱动，不同的数据库这个字段是不同的。下面是常见的数据库 ENGINE 写法。

```
django.db.backends.postgresql      #PostgreSQL
django.db.backends.mysql           #MySQL
django.db.backends.sqlite3         #SQLite
django.db.backends.oracle          #Oracle
```

(2) NAME：指定的数据库名，如果是 SQLite3，就需要填数据库文件的绝对路径，如 'NAME': os.path.join(BASE_DIR, 'db.sqlite3')。

(3) USER：数据库登录的用户名，MySQL 一般是 root。

(4) PASSWORD：登录数据库的密码，必须是 USER 用户对应的密码。

(5) HOST：由于一般的数据库都是 C/S 结构的，所以需要指定数据库服务器的位置，这里，数据库服务器和客户端都在一台主机上面，所以使用默认配置：127.0.0.1。

(6) PORT：数据库服务器端口，MySQL 的默认端口为 3306、Oracle 的默认端口为 1521、PostgreSQL 的默认端口为 5432。

注意：HOST 和 PORT 都不填时，使用的是默认配置，但是，如果要更改默认配置，就需要填入更改后的 HOST 和 PORT。

2.8.2 Django 的模型

接下来介绍如何利用 models.py 来建立一个数据库表，下面以用户 User 表为例。
models.py

```
from django.db import models
...
#Database
#https://docs.djangoproject.com/en/1.11/ref/settings/#databases

#用户
class User(models.Model):
    username = models.CharField(max_length=50)      #用户名
    password = models.CharField(max_length=50)      #密码
    email = models.EmailField()                     #Email

    def __str__(self):
        return self.username
...
```

User 表中包括 4 个字段。

(1) id：隐藏，主键，在 models.py 中不需要指明，从 1 开始的 int 类型（主键一般系统会自动添加，但是用户也可以自己定义，定义后在这个字段后面加上 primary_key=True）。

(2) username：用户名，models.CharField(max_length=50)，在后台维护界面表示为普通文本框，在数据库中表示 varchar(50)。

(3) password：密码，models.CharField(max_length=50)，在后台维护界面表示为普通文本框，在数据库中表示 varchar(50)。

(4) email：用户 Email 地址，models.EmailField()，在后台维护界面表示为 HTML5 格式的 Email 类型，在数据库中表现为 varchar(254)。

最后的 def __str__(self)方法表示如何将对象以 String 形式显示出来。

下面再来看收货地址的表结构。大家都知道，一个用户可以有多个收货地址，但是一个收货地址只对应一个用户，所以，用户与收货地址的关系是一对多的关系，这样就需要在地址表中建立一个外键，具体代码如下。

models.py

```
…
#收货地址
class Address(models.Model):
    user = models.ForeignKey(User)                    #关联用户id
    address = models.CharField(max_length=50)         #地址
    phone = models.CharField(max_length=15)           #电话

    def __str__(self):
        return self.address
…
```

这里，user = models.ForeignKey(User)，显而易见，建立了一个外键，在数据库中将会产生一个名为 user_id 的字段名，它与 User 表中的 id 字段对应。下面介绍 Django 中常见的字段类型，见表 2-2。

表 2-2　Django 中常见的字段类型

字　段	解　释	参数（*为必选）
AutoField	一个自动递增的整型字段，添加记录时它会自动增长	
BooleanField	布尔字段	
BinaryField	原始二进制字段	
BigIntegerField	大整数字段	
TextField	一个容量很大的文本字段	
CommaSeparatedIntegerField	用于存放逗号分隔的整数值	* max_length

ChoiceField	设置下拉菜单（即＜select＞标签），后面必须以 choice 参数指定的一个二维列表。例如：
	``
	class Foo(models.Model):
	    GENDER_CHOICES = (
	        ('M', 'Homme'),
	        ('F', 'Femme'),
	    )
	    gender = models.CharField(max_length=1,
	    choices=GENDER_CHOICES)
	``	

续表

字　　段	解　　释	参数(*为必选)
CharField	字符串字段	* max_length
DateField	日期字段	auto_now auto_now_add
DateTimeField	日期时间字段	auto_now auto_now_add
DecimalField	小数字段	
EmailField	Email 字段	
FileField	文件上传字段	* upload_to
FilePathField	指定目录上传文件	* path match base filename recursive
FloatField	浮点型字段	* max_digits * decimal_places
ImageField	类似 FileField,要校验上传的对象是否是一个合法图片	height_field width_field,
IntegerField	整数类型	
IPAddressField	字符串形式的 IP 地址	
NullBooleanField	允许 Null 的布尔字段	
PhoneNumberField	合法美国风格电话号码校验(格式：XXX-XXX-XXXX)	
PositiveIntegerField	无符号正整数字段,0 可以	
PositiveSmallIntegerField	正小整型字段	
SlugField	slug 是某个东西的小小标记(短签),只包含字母、数字、下画线和连字符	* max_length
SmallIntegerField	小整型字段	
TimeField	时间字段	
URLField	URL 字段	verify_exists
USStateField	美国州名缩写	
XMLField	XML 字符字段	schema_path

建立完毕后,运行如下命令：

```
\ebusiness>python manage.py makemigrations goods
\ebusiness>python manage.py migrate
```

注意：如果是 MySQL,就需要先建立一个名为 project 的数据库。

接下来运行下面的命令：

```
\ebusiness>python manage.py sqlmigrate goods 0001
```

得到如下结果。

```
BEGIN;
--
-- Create model Address
--
CREATE TABLE "goods_address" ("id" integer NOT NULL PRIMARY KEY AUTOINCREMENT, "address" varchar(50) NOT NULL, "phone"
varchar(15) NOT NULL);
--
-- Create model Goods
--
CREATE TABLE "goods_goods" ("id" integer NOT NULL PRIMARY KEY AUTOINCREMENT, "name" varchar(100) NOT NULL, "price" real
NOT NULL, "picture" varchar(100) NOT NULL, "desc" text NOT NULL);
--
-- Create model Order
--
CREATE TABLE "goods_order" ("id" integer NOT NULL PRIMARY KEY AUTOINCREMENT, "count" integer NOT NULL, "goods_id" integer
NOT NULL REFERENCES "goods_goods" ("id"), "order_id" integer NOT NULL REFERENCES
"goods_address" ("id"));
--
-- Create model Pay
--
CREATE TABLE "goods_pay" ("id" integer NOT NULL PRIMARY KEY AUTOINCREMENT, "create
_time" datetime NOT NULL, "status" bool
NOT NULL, "address_id" integer NOT NULL REFERENCES "goods_address" ("id"));
--
-- Create model User
--
CREATE TABLE "goods_user" ("id" integer NOT NULL PRIMARY KEY AUTOINCREMENT, "username" varchar(50) NOT NULL, "password"
varchar(50) NOT NULL, "email" varchar(254) NOT NULL);
--
-- Add field user to order
--
ALTER TABLE "goods_order" RENAME TO "goods_order__old";
CREATE TABLE "goods_order" ("id" integer NOT NULL PRIMARY KEY AUTOINCREMENT, "count" integer NOT NULL, "goods_id" integer
NOT NULL REFERENCES "goods_goods" ("id"), "order_id" integer NOT NULL REFERENCES
"goods_address" ("id"), "user_id" integer
NOT NULL REFERENCES "goods_user" ("id"));
```

```
INSERT INTO "goods_order" ("count", "order_id", "user_id", "id", "goods_id")
SELECT "count", "order_id", NULL, "id",
"goods_id" FROM "goods_order__old";
DROP TABLE "goods_order__old";
CREATE INDEX "goods_pay_address_id_dd8a2806" ON "goods_pay" ("address_id");
CREATE INDEX "goods_order_goods_id_55ff9645" ON "goods_order" ("goods_id");
CREATE INDEX "goods_order_order_id_797115ae" ON "goods_order" ("order_id");
CREATE INDEX "goods_order_user_id_bd5a6274" ON "goods_order" ("user_id");
--
--Add field user to address
--
ALTER TABLE "goods_address" RENAME TO "goods_address__old";
CREATE TABLE "goods_address" ("id" integer NOT NULL PRIMARY KEY AUTOINCREMENT,
"address" varchar(50) NOT NULL, "phone"
varchar(15) NOT NULL, "user_id" integer NOT NULL REFERENCES "goods_user" ("id"));
INSERT INTO "goods_address" ("address", "phone", "user_id", "id") SELECT "address",
"phone", NULL, "id" FROM "goods_address__old";
DROP TABLE "goods_address__old";
CREATE INDEX "goods_address_user_id_f98a2118" ON "goods_address" ("user_id");
COMMIT;
```

可以看出,这里都是一些初始化数据库的基本命令,Django 正是通过这个方法,即使开发人员不会 SQL 语句,也可以很好地开发 Django 网站。只要先修改 model.py 文件,然后执行上述两条命令,在.\goods\migrations 目录下就会新产生一个类似 0002_auto_20170820_1748.py 的文件,其中 0002 是序号,20170820_1748 表示在 2017 年 08 月 20 日 17 时 48 分建立,也可以通过\ebusiness＞python manage.py sqlmigrate goods 0002 命令查看对应的变化。

```
PS C:\Users\Jerry\ebusiness>python manage.py sqlmigrate goods 0002
BEGIN;
--
--Rename model Pay to Orders
--
ALTER TABLE "goods_pay" RENAME TO "goods_orders";
COMMIT;
PS C:\Users\Jerry\ebusiness>
```

登录后台后,可以看见这些表的字段,并且也可以用数据库管理工具进行查看。在这里可以看到这些表名都是为"应用名(goods)＋模块名＜models＞"定义的变量名,如 goods_order。

2.8.3 Django 的后台管理

经过上述操作,打开浏览器,输入 http://127.0.0.1:8000/admin/,用本书 2.6.3 节注册的超级管理员账号登录,就可以进入系统的后台。在这里可以对所建立表中的内容进行增加、删除、修改和查询（CRUD）操作。图 2-9 是刚才建立 address 表后台的管理界面。

图 2-9　建立 address 表后台的管理界面

单击 ➕ Add 链接，可以添加 address 表记录，如图 2-10 所示。

图 2-10　通过后台添加 address 表记录

单击 ✎ Change 链接，可以修改或删除 address 表记录，如图 2-11 所示。

图 2-11　通过后台修改或者删除 address 表记录

选择某条记录，出现如图 2-12 所示的界面，修改字段内容，然后单击 SAVE 按钮，对修改的记录进行保存操作。

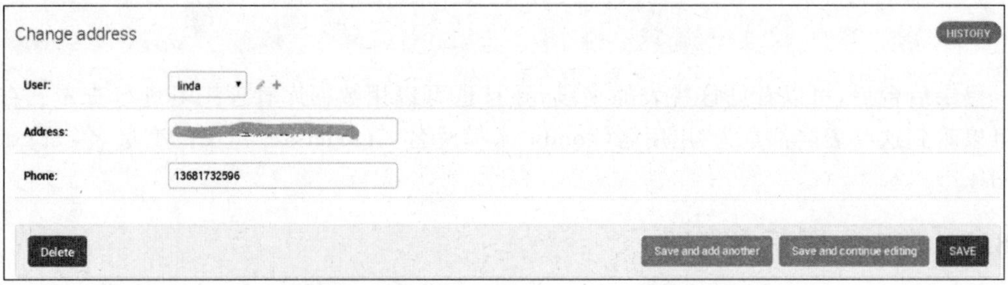

图 2-12　通过后台修改或者删除一条 address 表记录

或者单击 [Delete] 按钮,对记录进行删除操作。

还可以在图2-11上选择多条记录,然后在 Action 后的下拉列表框中选择 [Delete selected addresss ▼],单击 [Go] 按钮,删除这些记录,如图2-13所示。

图 2-13　通过后台删除一批 address 表中的记录

2.8.4　Django 如何对数据库进行操作

Django 对数据库的操作是在 views.py 中进行编码的。

1. 增加

```
变量 = 模板对象()
变量.对象变量1 = 值1
变量.对象变量2 = 值2
…
变量.对象变量n = 值n
变量.save()
```

例如:

```
…
    user = User()
    user.username = 'Peter'
    user.password = '123456'
    user.email = ' Peter_M@126.com '
    user.save()
…
```

把用户名为 Peter,密码为 123456,Email 为 Peter_M@126.com 的用户保存在数据库中。

2. 删除

```
模板对象.objects.filter(条件).delete()
```

例如：

```
…
Address.objects.filter(id=address_id).delete()
…
```

把 id 为 address_id 的记录从数据库表 Address 中删除。对于 objects.filter()方法,将在查询中进行介绍。

3. 修改

```
模板对象.objects.filter(条件).update(变量1=值1,变量2=值2,…,变量n=值n)
```

例如：

```
…
Address.objects.filter(id=address_id).update(address ='江苏省南京市秦淮区乌衣巷23号',phone ='13698763423')
…
```

把 id 为 address_id 的 Address 表记录中的地址改为："江苏省南京市秦淮区乌衣巷23号",电话号码改为："13698763423"。

4. 查询

1) 获取所有记录

```
变量=模板对象.objects.all()
```

这个方法返回的是对象表中所有记录的 QuerySet 对象。例如：

```
…
address_all =Address.objects.all()
…
```

返回 Address 表中所有记录的 QuerySet 对象。

2) 获取满足条件的对象

```
变量=模板对象.objects.get(条件)
```

这个方法返回的是符合条件的 Model 对象,类型为列表。例如：

```
…
address =Address.objects.get (id='1')
…
```

返回 Address 表中 id 为 1 的记录。

```
变量=模板对象.objects.filter (条件)
```

filter 和 get 类似,但支持更强大的查询功能。例如:

```
...
address =Address.objects.filter (id='1')
...
```

同样也是返回 Address 表中 id 为 1 的记录。

```
变量=get_object_or_404(模板对象,条件)
```

调用 django 的 get()方法,如果查询对象不存在,就会抛出一个 DoesNotExist 的异常,改为调用 django get_object_or_404()方法,它会默认调用 django 的 get()方法,如果查询对象不存在,就会抛出一个 Http404 的页面,这样对用户比较友好。所以,一般在编码的时候很少使用 get()方法,而使用 get_object_or_404()或者 filter()方法较多。例如:

```
...
from django.shortcuts import get_object_or_404
...
address =get_object_or_404 (Address,id='1')
...
```

返回 Address 表中 id 为 1 的记录。如果记录不存在,就进入系统 404(网页没有发现)页面。

下面详细介绍 filter 的用法,见表 2-3。

表 2-3 filter 的用法

参 数	介 绍
__exact	精确等于 like 'aaa'
__iexact	精确等于(忽略大小写)ilike 'aaa'
__contains	包含 like '%aaa%'
__icontains	包含(忽略大小写)ilike '%aaa%'
__gt	大于
__gte	大于或等于
__lt	小于
__lte	小于或等于
__in	存在于一个 list 范围内
__startswith	以…开头
__istartswith	以…开头(忽略大小写)
__endswith	以…结尾
__iendswith	以…结尾(忽略大小写)

续表

参　　数	介　　绍
__range	在…范围内
__year	日期字段的年份
__month	日期字段的月份
__day	日期字段的日
__isnull	=True/False

由此可见,使用 Django 编写网站,如果编码者不懂 SQL 语句照样可以进行网站的开发。

5．Django 的数据库 ORM 所有函数及操作

Django 的数据库 ORM 所有函数及操作总结如下。

(1) python manage.py shell。

直接根据 Django 的环境变量进入 shell 命令行。

(2) models.User.objects.all()。

查找 User 下的所有内容。

(3) models.User.objects.last()。

查找 User 最后一个内容。

(4) my.Email = 'jerrygu625@126.com'。

设置对象的 Email 字段为'jerrygu625@126.com '。

(5) my.save()。

直接保存数据库内容。

(6) models.User.objects.create()。

创建数据。

(7) models.User.objects.filter(username='jerrygu625', Email='126')。

过滤字段查找,相当于 SQL 中的 where。

(8) models.User.object.first()。

显示查询到的第一个元素。

(9) models.User.objects.filter(Email__contains='126')。

在字段元素中加上__contains 代表 SQL 语句中的 like 模糊查询。

(10) models.User.objects.filter(Email__icontains='126')。

忽略大小写的迷糊查询,将__contains 变成__icontains 便是忽略大小写了。

(11) models.User.objects.filter(id__range(1,10))。

范围查找,查找 id 为 1~10 的所有数据。

(12) models.User.objects.filter(username__contains='Cindy').update(username='jerrygu625')。

批量修改数据,首先模糊查询到想要的数据,然后通过 update 修改这些数据。

(13) models.User.objects.filter(username__contains='Cindy').delete()。

批量删除数据,首先模糊查询到想要的数据,然后通过 delete 删除这些数据。

2.9 Django 的视图管理

2.9.1 urls.py 中路径的定义

Django 的视图管理主要集中在 goods 目录下的 views.py 中完成业务逻辑以及在 ebusiness 目录下的 urls.py 中定义路径。只要在 views.py 中定义了一个方法:def dir (request[,…]),就必须在 urls.py 中定义路径。

```
...
from goods import views      #导入 goods 应用 views 文件
...
urlpatterns = [
    url(r'^dir/$', views.dir),
    ...
]
```

这样系统中就存在一个 http://127.0.0.1:8000/dir/路径。除了定义成 def dir (request),还可以在路径中定义参数。例如下列几种情况。

(1) views.py 中定义了一个方法:def dir(request,id),并且在 urls.py 中定义路径: url(r'^dir /(? P<id>[0-9]+)/ $ ', views.dir),定义了一个名为 id 的 0~9 组成的数字变量,如 1、12 等。

(2) views.py 中定义了一个方法:def dir(request,username),并且在 urls.py 中定义路径:url(r'^dir /(? P<username>[a-z,A-Z]+)/ $ ', views.username),定义了一个名为 username 的大小写字母组成的变量。

这里介绍在 Django 程序中文件路径的问题,由于 Django 的路径是在 urls.py 中定义的,所以对文件的读取(如显示图片)不能按照传统的方法(如 ../images/1.jpg)表示。首先在与 goods 目录平行的路径下建立一个文件夹,如文件夹 images,然后在 urls.py 中输入:

```
...
from django.views import static
...
urlpatterns = [
    url(r'^images/(?P<path>.*)',static.serve,{'document_root':'C:\\Python35\\Scripts\\ebusiness\\images'}),
    ...
]
```

其中,C:\\Python35\\Scripts\\ebusiness\\images 为绝对路径。

注意:在 Windows 下路径用\\(两个"\",前面的"\"为转义)。最后在模板中这样使用:<img src=/images/1.jpg>(注意,不是<img src=images/1.jpg>,images 前必须有

字符"/"），这样图片文件就可以正确显示了。再把上面的代码改为

```
...
import os

BASE_DIR = os.path.dirname(os.path.dirname(os.path.abspath(__file__)))
...
urlpatterns = [
    ...
    url(r'^static/(?P<path>.*)',static.serve,{'document_root':os.path.join(BASE_DIR,'images')}),
    ...
]
```

在这里先引入 os 类，然后定义 BASE_DIR 系统所在的文件位置，最后通过 os.path.join(BASE_DIR,'images')把系统路径与图片路径合在一起，这样当系统的文件路径发生变化的时候，也不需要修改程序代码。同样，把 css 文件放在这个目录下，就可以在模板中用 <link href="/images/signin.css" rel="stylesheet"> 使用 signin.css 文件。

url.py 中的符号见表 2-4。

表 2-4　url.py 中的符号

符　　号	说　　明	例　　子
^	指定的字符或字符串，如果放在[]中,表示否定	^add_address
$	指定的终止符或字符串	[0-9]+)/$
.	任何一个字符都可以	
所有字母与数字(含"/")	对应原有的字符	
[...]	括号中的内容表示一个字符格式的设置	
\d	任何一个数字字符,相当于[0-9]	
\D	非数字字符,相当于[^0-9]	
\w	任何一个字母或数字字符,相当于[a-zA-Z0-9]	
\W	任何一个非字母或数字字符,相当于[^a-zA-Z0-9]	
?	前面一个字符可以重复出现 0 或 1 次	(?P<orders_id>…)/
*	前面一个字符可以重复出现 0 或多次	?P<path>.*
+	前面一个字符可以重复出现 1 或多次	(?P<sign>[0-9]+)
{m}	表示前一字符可以出现 m 次(m 为数字)	
{m,n}	表示前一字符可以出现 m 到 n 次(m<n,且 m、n 为数字)	
|	或,即两种格式任选一种	
(?P<name>...)	同上,参数名为 name	(?P<orders_id>…)/

2.9.2 方法中显示内容

由 views.py 中的方法处理业务逻辑,最后的结果必须以网页的形式展示给用户,有以下几种方式。

(1) 通过 return HttpResponse(str) 在页面中直接显示,这种方式适用于代码开发阶段的调试,而不太适合真正的产品开发。

(2) 通过 render_to_response(template[,dictionary][,context_instance][,mimetype]):调用模板 template,可选项为 dictionary、context_instance 和 mimetype,但是这个方法将被 render 逐步取代。

(3) 通过 render(request,template[,dictionary][,context_instance][,context_instance][,status][,current_app]),调用模板 template,可选项为 dictionary、context_instance、context_instance、status 和 current_app,是一个全新快捷的 render_to_response,Django 1.3 将开始使用。其中[,dictionary]经常被使用,目的是向 template 传输参数。

(4) 通过 response = HttpResponseRedirect('/other_dir/')...return response:转向目录/other_dir/。

注意:如果一个方法中任何一个分支没有上述四项中的一项,程序将报 500 号内部错误。

2.9.3 处理表单

Django 经常以如下方式处理表单。

```
...
def dir(request)
    ...
    if request.method =="POST":
        #如果是表单提交状态,就处理表单
        #如果调用定义表单,就构造表单变量
        uf =NameForm(request.POST)
        #验证表单信息是否正确
        if uf.is_valid():
            #获取表单信息
            var1 =request.POST.get(' var1')
            #不要用 var1=uf.cleaned_data['var1'],这样不便于做接口测试
            varN =request.POST.get('varN')
            #进行判断操作
            CheckValue =ObjectVar.objects.filter([Condition])
            if CheckValue:
                #如果满足条件,就进行操作
                ...
                #返回成功页面
                ...
            else:
```

```
            #否则进行错误操作
            ...
            #返回出错页面
            ...
    else:
        #如果不是表单提交状态,就显示表单
        #如果调用定义表单,就构造表单变量
        uf = NameForm ()
        #进入表单显示界面
        ...
...
```

表单请求流程如图 2-14 所示。

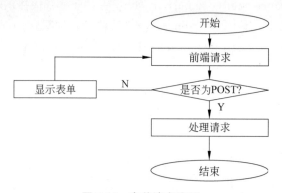

图 2-14　表单请求流程

页面接受请求,如果不是 POST 方式,就显示表单。当表单提交后,方式变为 POST,然后进行处理工作,如表单信息的获取、数据有效性验证、对数据库的操作等。

这样就可以用一个方法显示表单以及处理表单信息了。这里以用户注册为例。

```
...
#用户注册
def register(request):
    if request.method == "POST":
        uf = UserForm(request.POST)
        if uf.is_valid():
            #获取表单信息
            username = request.POST.get('username')
            password = request.POST.get('password')
            email = request.POST.get('email')
            #查找数据库中是否存在相同的用户名
            user_list = User.objects.filter(username=username)
            if user_list:
                return render_to_response('register.html',{'uf':uf,"error":"用户名已经存在!"})
```

```
        else:
            #将表单写入数据库
            user = User()
            user.username = username
            user.password = password
            user.email = email
            user.save()
            #返回注册成功页面
            uf = LoginForm()
            return render_to_response('index.html',{'uf':uf})
    else:
        uf = UserForm()
    return render_to_response('register.html',{'uf':uf})
...
```

当然,处理进入成功页面是在 if 中,还是在 else 中可以根据 if 后的判断条件决定。

2.9.4 分页功能

在显示列表中,经常会使用到分页的功能,Django 提供了比较易用的分页功能。在 views.py 中,代码如下。

```
...
from django.core.paginator import Paginator, EmptyPage, PageNotAnInteger
...
    paginator = Paginator(ListVar, 5) #ListVar 为需要分页的列表变量,5 为每页 5 条记录
    page = request.GET.get('page')#page 为页号
    try:
        contacts = paginator.page(page)
    except PageNotAnInteger:
        #如果页号不是整数,就转入第一页
        contacts = paginator.page(1)
    except EmptyPage:
        #如果页号超出最大页号,就转向最后一页
        contacts = paginator.page(paginator.num_pages)
    render(request,template[,dictionary][,context_instance][,context_instance]
[,status][, current_app])
...
```

然后在模板文件中这样应用。

```
...
<!--列表分页器-->
    <div class="pagination">
        <span class="step-links">
```

```
            {%if goodss.has_previous %}
                <a href="? page={{ goodss.previous_page_number }}">上一页</a>
            {%endif %}
            <span class="current">
                Page {{ goodss.number }} of {{ goodss.paginator.num_pages }}.
            </span>
            {%if goodss.has_next %}
                <a href="? page={{ goodss.next_page_number }}">下一页</a>
            {%endif %}
        </span>
    </div>
...
```

分页的界面请参看图3-12。

2.10 Django 的模板管理

处理完毕的数据通过 views.py 传入到模板文件中，文件模板通过标签把它显示出来，本节主要讨论这个问题。

2.10.1 变量的使用

{{var_name}}：var_name 为从 views.py 传过来的参数变量，当页面显示的时候显示的是参数的值。

```
<p style="color:red">{{error}}</p><br>
```

error 为从 views.py 传过来的参数变量，显示的时候显示的是变量 error 的值。

2.10.2 标签的使用

标签的使用是在模板中使用简单的程序控制变量输出，特别是变量不是 Python 的基本类型（如数字、字符串）时，下面对这些变量进行一一介绍。

(1){% if %}…{% endif %}：可以使用 and、or、not 组织逻辑，但不允许 and 和 or 同时出现，在条件语句中，Django 1.10 版本已经支持{% elif %}用法。

```
{%if age >18 %}
    ...
{%else %}
    ...
{%endif %}
```

(2){% ifchanged %}…{% endifchanged %}：检测本次循环的值和上一次循环的值是否一致，只能用在循环体里面。使用这个方法，如果是直接检测循环变量是否发生变化，使用方法如下。

```
{%ifchanged %}
...
{%endifchanged %}
```

如果检测循环变量的某个下级变量,如循环变量是 date,就检测 date.hour,使用如下代码。

```
{%ifchanged date.hour%}
...
{%endifchanged %}
```

ifchanged 也可以加上一个{% else %}语句。

(3){%firstof%}。

```
{%firstof var1 var2 var3%}
```

等价于

```
{%if var1 %}
        {{ var1 }}
    {%else %}
        {%if var2 %}
            {{ var2 }}
        {%else %}
            {%if var3 %}
                {{ var3 }}
            {%endif %}
        {%endif %}
{%endif %}
```

(4){% ifequal %}…{% endifequal %}:这个将被淘汰,不进行介绍。

(5){% ifnotequal% }…{% endifnotequal %}:这个也将被淘汰,不进行介绍。

(6){% for %}…{% endfor %}:用来循环一个列表,还可以使用 resersed 关键字进行倒序遍历,一般可以使用 if 语句先判断列表是否为空,然后再进行遍历,还可以使用 empty 关键字判断,参见(7)。

for 标签中可以使用 forloop。

① forloop.counter:当前循环计数,从 1 开始。

② forloop.counter0:当前循环计数,从 0 开始,标准索引方式。

③ forloop.revcounter:当前循环的倒数计数,从列表长度开始。

④ forloop.revcounter0:当前循环的倒数计数,从列表长度减 1 开始。

⑤ forloop.first bool 值:判断是不是循环的第一个元素。

⑥ forloop.last bool 值:判断是不是循环的最后一个元素。

⑦ forloop.parentloop:用在嵌套循环中,得到 parent 循环的引用,然后可以使用以上参数。

```
{%for key in list %}
    ...
{%key %}
    ...
{%endfor %}
```

for 除了可以循环列表型变量,也可以循环字典型变量,例如:

```
{%for key,value in dict %}
    ...
{%key %}{%value %}
    ...
{%endfor %}
```

(7) {% for %}…{% empty %}…{% endfor %}:当 for 变量为空的时候,能够执行 empty 后的内容。

```
{%for variable in list %}
    ...
{%empty %}
    ...
{%endfor %}
```

其形式等同于先判断 list 是否存在,然后再根据情况做相应的操作。

(8) {% cycle %}:在循环时轮流使用给定字符串列表中的值。

```
{%for key,value in dict %}
    ...
<tr bgcolor="{%cycle '#CECECE' '#ECECEC'%}">...</tr>
    ...
{%endfor %}
```

(9) {#…#}:单行注释。{% comment %}…{% endcomment %}:多行注释。

```
{%comment %}
    ...
{%endcomment %}
    ...
{%#...#%}
```

(10) {% csrf_token %}:生成 csrf_token 的标签,用于防止 CSRF 攻击的验证,本书将在 4.2 节中对其进行详细介绍。

(11) {% filter %}…{% endfilter %}:将 filter 标签圈定的内容执行过滤器操作。关于过滤器,参见 2.10.3 节。

```
{%filter force_escape|lower%}
...
{%endfilter%}
```

(12) `{% load %}`：加载标签库。例如，定义一个标签库文件，如 exists_filter.py，内容如下。

```
from django import template

register = template.Library()
@register.filter(name='exists')
def exists(tool,tool_cfgs):
    return True if tool in tool_cfgs else False
```

在模板中调用以下代码。

```
{%load exists_filter%}

{%if 'viewstyle'|exists:tool_cfgs %}
    //todo something
{%endif %}
```

(13) `{% now %}`：获取当前时间。

```
{%now %}
```

如果需要转义，则调用如下代码。

```
{%now "jS o\f F" %}
```

因为 f 是格式化字符。具体的格式化时间日期格式的字符串如表 2-5 所示。

表 2-5 格式化时间日期格式的字符串

符号	解 释	例 子
a	'a.m.'或者 'p.m.'（注意，这与 PHP 的输出略有不同，因为这包括与相应的新闻样式相匹配的周期）	'a.m.'、'p.m.'
A	'AM'或者'PM'	'AM'、'PM'
b	三个字母组成小写缩略格式的月	'jan'
d	两位带前面 0 填充月的数字	'01'~'31'
D	三个字母的缩写周	'Fri'
f	分钟，如果后面是:00，则忽略不显示	'1'、'1:30'
F	长格式的月	'January'
g	没有前面 0 填充 12 进制的小时	'1'~'12'

续表

符号	解 释	例 子
G	没有前面0填充二十四进制的小时	'0'~'23'
h	十二进制的小时	'01'~'12'
H	二十四进制的小时	'00'~'23'
i	分钟	'00'~'59'
j	没有前面0填充一个月中的天	'1'~'31'
l	长格式的周	'Friday'
L	是否有闰年,布尔格式	True 或 False
m	用两位数字表示月	'01'~'12'
M	由3个字母组成的缩写的月,第一个字母为大写,后面的字母为小写,如'Jan'。而符号 b 表示3个字母都为小写,如'jan'	'Jan'
n	没有前面0填充的月	'1'~'12'
N	上一个月,按照新闻格式,自由扩展	'Jan. '、'Feb. '、'March'、'May'
O	格林尼治时间的差异	'+0200'
P	时间,格式为"小时[:分钟] a. m. /p. m."(其中,":分钟"可以没有,如果是0点,需要表示成 midnight 或 noon)	'1 a. m. '、'1:30 p. m. '、'midnight'、'noon'、'12:30 p. m. '
r	RFC 2822 格式日期	'Thu, 21 Dec 2000 16:01:07+0200'
s	没有前面0填充的秒	'00'~'59'
S	2个字符的英语字母顺序后缀	'st'、'nd'、'rd'或'th'
t	制定月份中的天数	28~31
T	这台机器的时区	'EST'、'MDT'
w	从0开始的一周中天数	'0' (Sunday) ~'6' (Saturday)
W	每年 ISO-8601 周数,本周从星期一开始	1~53
y	两位数字的年	'17'
Y	四位数字的年	'2017'
x	一年中的第几天	0~365
X	秒的时区偏移量。UTC 西边的时区偏移量总是负的,UTC 东边的时区偏移量总是正的	

(14) {% url %}:给定某个 module 中方法的名字,给定参数,那么模板引擎产生一个 URL,从而避免硬编码 URL 到代码中。

注意:前提是 URLconf 中存在相应的映射,如果 URLconf 中没有该映射,就会抛出异常。

```
{%url path.to.view arg1 ,arg2 as the url %}
<a href="{{ the_url }}">Link to optional stuff</a>
```

这相当于

```
{%url path.to.view as the_url %}
{%if the_url %}
<a href="{{ the_url }}">Link to optional stuff</a>
{%endif %}
```

(15) {% verbatim %}…{% endverbatim %}：禁止 render(渲染)。

```
{%verbatim %}
…
{%endverbatim %}
```

(16) {% with %}：当一个变量访问消耗很大的时候，可以用另外一个变量替换它，这种替换只在 with 内部有效。

```
{%if age >18 %}
    {%with patient as p %}
{%else %}
    {%with patient.parent as p %}
    …
{%endwith %}
{%endif %}
```

(17) {% autoescape %}…{% endautoescape %}：自动转义 HTML 元素。如果不愿意自动转义，可以通过以下方式关闭。

```
{%autoescape off %}
…
{%endautoescape %}
```

另外，也可以通过过滤器关闭自动转义。

```
{{ data|safe }}
```

(18) {% extends %}：表示本模板要对指定的父模板进行扩展。

```
{%extends "AAA.html" %}
```

或者扩展对象是一个字符串变量。

```
{%extends variable %}
```

(19) {% block %}…{% endblock %}：定义一个块。在 base.html 中如下定义。

```
{%block content %}

{%endblock %}
```

在 index.html 中有如下所示这样的代码。

```
{%extends "base.html"%}
{%block content%}
...
{%endblock%}
```

这样,index.html 中{% block content %}…{% endblock %}中间内容就嵌入到 base.html 的{% block content %}…{% endblock %}之间。

(20){% include %}：将另外一个模板文件中的内容添加到该文件中。注意区别 extend 是继承。

```
{%include "foo/bar.html"%}
```

或者

```
{%include template_name%}
```

(21){% spaceless %}…{% endspaceless %}：删除包围内容中的所有 tab 或者回车字符。

```
{%spaceless%}
...
{%endspaceless%}
```

(22){% templatetag %}：模板系统本身没有转义的概念,因此,如果要输出一个像{%这样的标签,就需要采用这种方式,否则就会出现语法错误。

参数有如下几个。
① openblock——>{%。
② closeblock——>%}。
③ openvariable——>{{。
④ closevariable——>}}。
⑤ openbrace——>{。
⑥ closebrace——>}。
⑦ opencomment——>{#。
⑧ closecomment——>#}。

```
{%templatetag%}
```

2.10.3 过滤器的使用

先看一个例子,{{ ship_date|date : "F j, Y" }},ship_date 变量传给 data 过滤器,data 过滤器通过使用"F j, Y"这几个参数格式化日期数据。"|"代表类似 UNIX 命令中的管道操作。其中,过滤参数同{%now%}。

虽然模板中提供了各种语句,但是在真正的工作中,对业务逻辑的处理,还是在 views.py 中进行,而模板中的语句仅仅用于对数据的显示,否则就违背了 MVC 的设计初衷。

Django 的常用过滤器见表 2-6。

表 2-6　Django 的常用过滤器

过滤器	解　　释	案　　例
add	给变量加上相应的值	{{ user.age | add:"5" }}　# 空格不要乱加
addslashes	在变量中的引号(如双引号、单引号)前加上斜线	
capfirst	第一个字母大写	{{"good"| capfirst }} 返回"Good"
center	输出指定长度的字符串,把变量居中	{{ "abcd"| center:"50" }}
cut	从字符串中移除指定的字符	{{ "You are not a Englishman" | cut:"not"}}
date	格式化日期字符串	
default	如果值不存在,则使用默认值代替	{{ value | default:"(N/A)" }}
default_if_none	如果值为 None,则使用默认值代替	
dictsort	按某字段排序,变量必须是一个 dictionary	{% for moment in moments | dictsort:"id" %}
dictsorted	按某字段倒序排列,变量必须是 dictionary	
divisbleby	判断是否可以被数字整除	{{ 224 | divisibleby:2 }} 返回 True
escape	按 HTML 转义,如将"<"转换为"&lt"	
escapejs	替换 value 中的某些字符,以适应 JavaScript 和 Json 格式	
filesizeformat	增加数字的可读性,转换结果为 13KB、89MB、3Bytes 等	{{ 1024 | filesizeformat }} 返回 1.0KB
first	返回列表中的第一个值,变量必须是一个列表	
floatformat	转换为指定精度的小数,默认保留 1 位小数	{{ 3.1415926 | floatformat:3 }}返回 3.142,四舍五入
get_digit	从个位数开始截取指定位置的数字	{{ 123456 | get_digit:'1'}}
join	用指定分隔符连接列表	{{ ["abc","45"] | join:" * "}}返回 abc * 45
last	返回列表中的最后一个值,变量必须是一个列表	
length	求字符串或者列表的长度	
length_is	比较字符串或者列表的长度	{{ 'hello'| length_is:'3' }}

续表

过滤器	解 释	案 例
linebreaks	用\<p\>或\<br\>标签包裹变量	{{ "Hi\n\nDavid"|linebreaks }} 返回\<p\>Hi\</p\>\<p\>David\</p\>
linebreaksbr	用\<br/\>标签代替换行符	
linenumbers	为变量中的每行加上行号	
ljust	输出指定长度的字符串,变量左对齐	{{"ab"|ljust:5}}返回 "ab"
make_list	将字符串转换为列表	
pluralize	根据数字确定是否输出英文复数符号	
random	返回列表的随机一项	
removetags	删除字符串中指定的 HTML 标记	{{value | removetags:"h1 h2"}}
rjust	输出指定长度的字符串,变量右对齐	
safe	对某个变量关闭自动转义	{{ value|safe }}
slice	切片操作,返回列表	{{[3,9,1] | slice:":2"}}返回[3,9] {{"asdikfjhihgie"|slice:':5'}}返回 "asdik"
slugify	在字符串中留下减号和下画线,其他符号删除,空格用减号替换	{{'5-2=3and52=3'|slugify}}返回 5-23and5-23
stringformat	字符串格式化,语法同 Python	
striptags	过滤掉 html 标签	
time	返回日期的时间部分	
timesince	以"到现在为止过了多长时间"显示时间变量	结果可能为 45days, 3 hours
timeuntil	以"从现在开始到时间变量"还有多长时间显示时间变量	
title	每个单词首字母大写	
truncatechars	按照字符截取字符串	{{ value|truncatechars:5 }}
truncatewords	将字符串转换为省略表达方式	{{'This is a pen'| truncatewords:2 }}返回 This is …
truncatewords_html	同 truncatewords,但保留其中的 HTML 标签	{{ '\<p\>This is a pen\</p\>' | truncatewords:2 }}返回 \<p\>This is ...\</p\>
turncatewords	按照单词截取字符串(其实就是按照空格截取)	
upper\lower	以大\小写方式输出	{{ user.name | upper }}
urlencode	将字符串中的特殊字符转换为 URL 兼容表达方式	{{'http://www.aaa.com/foo?a=b&b=c'|urlencode}}

续表

过滤器	解　释	案　例
urlize	将变量字符串中的 URL 由纯文本变为链接	
wordcount	返回变量字符串中的单词数	
yesno	将布尔变量转换为字符串 yes，no 或 maybe	{{ True | yesno }} {{ False | yesno }} {{ None | yesno }} 返回 yes no maybe

2.11 基于 Python Requests 类数据驱动的 HTTP 接口测试

2.11.1 测试金字塔

图 2-15 是 Main Cohn 提出的软件测试金字塔，他认为测试工程师应该把大量的工作花在单元测试和接口测试方面，其余工作花在 UI 测试以及探索式测试方面。纵然，单元测试的优点很突出，它接近代码本身，执行速度快，开发者可以一边写产品代码，一边写单元测试代码，一旦在单元测试中发现缺陷，就可以马上找到对应的产品代码进行修改。然而，单元测试的缺点也很明显，就是有多少产品代码，就要有相应的单元测试代码与它对应，这样造成的结果是单元测试代码等于甚至超过产品代码的数量，这也就是为什么单元测试在一般的中小型企业很难全面被推广的原因。对于基于 UI 层面的测试，由于需求变更，页面调整比较频繁，所以，在许多企业基于 UI 自动化测试仅用于需求变化不大的核心功能的自动化，往往是一些冒烟测试用例。而基于两者之间的接口测试（Interface Test），牵扯到的代码比单元测试要少得多，并且基本上不受页面变更的影响，所以越来越受到广大软件开发者的喜爱。

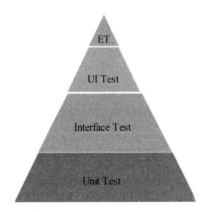

图 2-15　软件测试金字塔

2.11.2 unittest

由于本书是介绍 Django，而 Django 是基于 Python 语言的，所以接下来主要介绍基于 Python Requests 类的软件接口测试。首先介绍基于 Python 的单元测试框架 unittest，unittest 原名为 pytest，它是属于 XUnit 框架下的。这里先来看一段产品代码。

Calculator.py：

```python
#!/usr/bin/env python
#coding:utf-8

class calculator:
    def __init__(self, a, b):
        self.a=int(a)
        self.b=int(b)

    def myadd(self):
        return self.a+self.b

    def mysubs(self):
        return self.a-self.b

    def mymultiply(self):
        return self.a*self.b

    def mydivide(self):
        try:
            return self.a/self.b
        except ZeroDivisionError:
            print ("除数不能为零")
            return 0
```

很显然，这个代码具有实现加、减、乘、除四则运算的功能。类 calculator 有两个成员变量：self.a 和 self.b。方法 myadd、mysubs、mymultiply 和 mydivide 分别实现 self.a 与 self.b 的加、减、乘、除四个功能，即 self.a+self.b、self.a-self.b、self.a*self.b 和 self.a/self.b。在 mydivide 中，如果除数 self.b 为 0，就进行对应的处理，打印"除数不能为零"的警告，然后返回 0。现在观察这段代码对应的 unittest 框架的测试代码。

CalculatorTest.py：

```python
#!/usr/bin/env python
#coding:utf-8
import unittest
from Calculator import calculator
```

```python
class calculatortest(unittest.TestCase):
        def setUp(self):
                print("Test start!")

        def test_base(self):
                j=calculator(4,2)
                self.assertEqual(j.myadd(),6)
                self.assertEqual(j.mysubs(),2)
                self.assertEqual(j.mymultiply(),8)
                self.assertEqual(j.mydivide(),2)

        def test_divide(self):
                j=calculator(4,0)
                self.assertEqual(j.mydivide(),0)

        def tearDown(self):
                print("Test end!")

if __name__=='__main__':
        #构造测试集
        suite=unittest.TestSuite()
        suite.addTest(calculatortest("test_base"))
        suite.addTest(calculatortest("test_divide"))
        #运行测试集合
        runner=unittest.TextTestRunner()
        runner.run(suite)
```

(1) 首先使用 unittest 测试框架必须先引入 unittest 类：import unittest，unittest 类是 Python 默认的自带测试类，只要安装了 Python，这个类就自动安装上了。

(2) 然后引入被测试类：from Calculator import calculator。

(3) unittest 的测试方法的类参数必须为 unittest.TestCase，即 class calculatortest (unittest.TestCase):。

(4) 和其他 XUnit 测试框架一样，unittest 也存在一个初始化方法和一个清除方法，分别定义为 def setUp(self):和 def tearDown(self):，由于这里没有具体实际性的操作，仅在 def setUp(self):方法中打印一个"Test start!"字符串；在 def tearDown(self):方法中打印一个"Test end!"字符串，以标识测试程序的开始与结束。

(5) unittest 具体测试方法的方法名必须以 test_开头，这有点类似 JUnit3。语句 j= calculator(4,2)先定义一个 self.a＝4 和 self.b＝2 的类变量 j，然后通过断言 self.assertEqual(操作值，期待值)方法验证计算结果与预期结果是否一致。

(6) 在 def test_divide(self):方法中专门对除数为 0 的情况进行测试。

(7) unittest 的主方法与其他主方法一样，为 if __name__＝＝'__main__':，先通过 suite ＝unittest.TestSuite()构造测试集，然后通过 suite.addTest(calculatortest("test_base"))，suite.addTest(calculatortest("test_divide"))把两个测试方法加进去，接下来通过 runner＝

unittest.TextTestRunner()和 runner.run(suite)执行测试工作。

下面简单介绍一下 unittest 中所有的断言方法,见表 2-7。

表 2-7　unittest 中所有的断言方法

序号	断言方法	断言描述
1	assertEqual(arg1,arg2,msg=None)	验证 arg1=arg2,若不等,则返回 Fail
2	assertNotEqual(arg1,arg2,msg=None)	验证 arg1!= arg2,若相等,则返回 Fail
3	assertTrue(expr,msg=None)	验证 expr 是 True,如果为 False,则返回 Fail
4	assertFalse(expr,msg=None)	验证 expr 是 False,如果为 True,则返回 Fail
5	assertIs(arg1,arg2,msg=None)	验证 arg1、arg2 是同一个对象,若不是,则返回 Fail
6	assertIsNot(arg1,arg2,msg=None)	验证 arg1、arg2 不是同一个对象,若是,则返回 Fail
7	assertIsNone(expr,msg=None)	验证 expr 是 None,若不是,则返回 Fail
8	assertIsNotNone(expr,msg=None)	验证 expr 不是 None,若是,则返回 Fail
9	assertIn(arg1,arg2,msg=None)	验证 arg1 是 arg2 的子串,若不是,则返回 Fail
10	assertNotIn(arg1,arg2,msg=None)	验证 arg1 不是 arg2 的子串,若是,则返回 Fail
11	assertIsInstance(obj,cls,msg=None)	验证 obj 是 cls 的实例,若不是,则返回 Fail
12	assertNotIsInstance(obj,cls,msg=None)	验证 obj 不是 cls 的实例,若是,则返回 Fail

当许多测试代码需要批量运行的时候,可以进行如下操作。

(1) 把这些测试代码的文件名定义成一个可以用正则方法匹配的模式,例如,都以 Test 结尾的.py 文件:pattern="*Test.py"。

(2) 建立一个 py 文件,如 runtest.py。

```python
#!/usr/bin/env python
#coding:utf-8
import unittest

test_dir='./'
discover=unittest.defaultTestLoader.discover(test_dir,pattern="*Test.py")

if __name__=='__main__':
    runner=unittest.TextTestRunner()
    runner.run(discover)
```

(3) test_dir='./':定义测试代码的路径,这里为当前路径。

(4) discover=unittest.defaultTestLoader.discover(test_dir,pattern="*Test.py"):为调用测试路径下以 Test 结尾的.py 文件(pattern="*Test.py")。

(5) 在主方法中通过调用 runner = unittest.TextTestRunner()和 runner.run(discover)两行代码实现匹配的所有测试文件中测试用例的执行。

既然介绍到 unittest 的批量操作,很有必要介绍一下如何通过 unittest 生成一份好看的测试报告。

读者可以先到网站 http://tungwaiyip.info/software/HTMLTestRunner.html 下载 HTMLTestRunner.py 文件到%PYTHON_HOME%\Lib\目录下。如果使用的是 Python 2.x 系列,就不需要进行修改了;如果使用的是 Python 3.x 系列,请作如下修改。

```
94 行
import StringIO
改为
import io

539 行
self.outputBuffer=StringIO.StringIO()
改为
self.outputBuffer=io.StringIO()

631 行
print >>sys.stderr, '\nTime Elapsed: %s' % (self.stopTime-self.startTime)
改为
print (sys.stderr, '\nTime Elapsed: %s' % (self.stopTime-self.startTime))

642 行
if not rmap.has_key(cls):
改为
if not cls in rmap:

766 行
uo =o.decode('latin-1')
改为
uo =o

772 行
ue =e.decode('latin-1')
改为
ue =e
```

这样,在 runtest.py 头部加入 from HTMLTestRunner import HTMLTestRunner,然后在 runner.run(discover) 前面加上 fp = open("result.html","wb"), runner = HTMLTestRunner(stream=fp,title='测试报告',description='测试用例执行报告'),后面加上 fp.close(),运行测试用例完毕,就可以生成一份美观的基于 HTML 的测试报告了。最后的 runtest.py 代码如下。

```
#!/usr/bin/env python
#coding:utf-8
import unittest
```

```python
from HTMLTestRunner import HTMLTestRunner

test_dir='./'
discover=unittest.defaultTestLoader.discover(test_dir,pattern="*Test.py")

if __name__=='__main__':
    runner=unittest.TextTestRunner()
    #以下用于生成测试报告
    fp=open("result.html","wb")
    runner =HTMLTestRunner(stream=fp,title='测试报告',description='测试用例执行报告')
    runner.run(discover)
    fp.close()
```

图 2-16 是基于 HTML 的 unittest 测试报表,这里的测试用例比上面介绍的要多一些。

测试报告

Start Time: 2017-08-29 15:08:09
Duration: 0:00:00.014037
Status: Pass 4

测试用例执行报告

Show Summary Failed All

Test Group/Test case	Count	Pass	Fail	Error	View
CalculatorTest.calculatortest	4	4	0	0	Detail
test_base			pass		
test_divide			pass		
test_multiply			pass		
test_subs			pass		
Total	4	4	0	0	

图 2-16 unittest 测试报表

2.11.3　requests 对象的介绍与使用

requests 对象是用 Python 语言编写的,它基于 urllib,采用的是 Apache2 Licensed 开源协议的 HTTP 库。但是,它比 urllib 更方便,可以减少大量的工作,并且它完全满足 HTTP 测试的需求。Requests 的哲学是:以 PEP 20[①] 的习惯用语为中心开发的,所以它比 urllib 更加符合 Python 思想。另外,更重要的一点是它支持 Python 3.x 系列。

要是用 requests 进行接口测试,首先要下载 requests 类,可以用老办法实现,即通过 pip 命令下载。

```
>pip install requests
```

还可以使用下面的方法安装。

```
>git clone git://github.com/kennethreitz/requests.git
```

---

① PEP 20 就是本书开始提到的 Python 禅歌。

```
>cd requests
>python setup.py install
```

下面介绍 requests 对象的使用。

(1) 通过 requests 发送 GET 请求。

```
response = requests.get(url,params= payload)
```

url 为发送的地址，payload 为请求的参数，格式为字典类型，前面变量名必须为 params，response 为返回的变量。

例如：

```
url ="http://www.a.com/user.jsp"
payload={"id":"1","name":"Tom"}
data = requests.get(url,params=payload)
```

(2) 通过 requests 发送 POST 请求。

```
response = requests.post(url,data=payload)
```

url 为发送的地址，payload 为请求的参数，格式为字典类型，前面变量名必须为 data，response 为返回的变量。

例如：

```
url ="http://www.b.com/login.jsp"
payload={"username":"Tom","password":"123456"}
data = requests.post(url,data=payload)
```

POST 除了可以支持参数，也可以支持 JSON(JS 对象标记)，方法如下。

```
import json
response = requests.post('http://www.a.com/path/', data = json.dumps ({'name':
'Jerry'}))
print(r.json())
```

下面的方法可以定制化 head。

```
import json
data ={'name': 'Jerry'}
headers ={'content-type': 'application/json',
         'User-Agent':'Mozilla/5.0 (X11;Ubuntu;Linuxx86_64;rv:22.0) Gecko/
20100101 Firefox/22.0'}
response = requests.post(' http://www.a.com/path/', data=data, headers=headers)
```

(3) 使用 PUT、DELETE、HEAD 和 OPTIONS 发送请求。

虽然这几种方法用得不多，但是为了完整性，这里对它们进行简单介绍。

```
requests.put("http://www.b.com/put")
requests.delete("http://www.b.com/delete")
requests.head("http://www.b.com/get")
requests.options("http://www.b.com/get")
```

(4) requests 的返回值

requests 的返回值的类型为 HttpResponse 对象,其存储了服务器响应的内容,这里对其进行介绍。见表 2-8,这里假设 requests 的返回值为 response。

表 2-8 requests 的返回值

编号	代　　码	解　　释
1	response.status_code	返回应答消息中的返回状态码
2	response.raw	返回原始的响应体,也就是 urllib 的 response 对象,使用 r.raw.read() 读取
3	response.content	返回字节方式的响应体,会自动解码 gzip 和 deflate 压缩
4	response.headers	返回以字典对象存储服务器响应头,但是这个字典比较特殊,字典键不区分大小写,若键不存在,则返回 None
5	response.cookies	返回网址的 cookies 信息
6	response.url	返回网址的地址
7	response.history	返回的历史记录(以列表形式显示)
8	response.text	返回网址的内容信息
9	response.json()	返回 Requests 中内置的 JSON 解码器
10	response.raise_for_status()	失败请求(非 200 响应)抛出异常

1.3 节介绍了 HTTP 的返回码,但是经常使用的仅有以下几个。

① 200(OK):客户端请求成功。

② 304(No Changed):没有改变。

③ 401(Unauthorized):请求未授权,这个状态代码必须和 WWW-Authenticate 报文域一起使用。

④ 400(Bad Request):客户端请求有语法错误,不能被服务器端理解。

⑤ 403(Forbidden):服务器请求被收到,但是拒绝提供服务。

⑥ 404(Not Found):请求资源不在,例如,错误的 URL。

⑦ 500(Internal Server Error):服务器内部错误。

⑧ 503(Server Unavailable):服务器当前不能处理客户请求,一段时间后可能恢复正常。

有了上面这些知识,下面介绍如何通过 requests 类实现接口测试,这里以前面介绍的登录模块作为测试对象设置测试用例。登录模块测试用例见表 2-9。

表 2-9　登录模块测试用例

编号	描　　述		期　望　结　果
	用户名	密码	
1	正确	错误	有提示信息"用户名或者密码错误"
2	错误	正确	有提示信息"用户名或者密码错误"
3	错误	错误	有提示信息"用户名和密码错误"
4	正确	正确	登录后进入页面,出现"查看购物车"

假设这里正确用户名为 jerry,正确密码为 123456,设计测试代码如下。

testLogin.py:

```
import requests

correctusername="jerry"
correctpassword="123456"
discorrectusername="cindy"
discorrectpassword="000000"

url="http://localhost:8080/sec/20/jsp/index.jsp"
#正确的用户名,错误的密码
payload={"name":correctusername,"password":discorrectpassword}
data = requests.post(url,data=payload)
if (str(data.status_code)=="200") and ("用户名或者密码错误" in str(data.text)):
    print("pass")
else:
    print("Fail")
#错误的用户名,正确的密码
payload={"name":discorrectusername,"password":correctpassword}
data = requests.post(url,data=payload)
if (str(data.status_code)=="200") and ("用户名或者密码错误" in str(data.text)):
    print("pass")
else:
    print("Fail")
#错误的用户名,错误的密码
payload={"name":discorrectusername,"password":discorrectpassword}
data = requests.post(url,data=payload)
if (str(data.status_code)=="200") and ("用户名和密码错误" in str(data.text)):
    print("pass")
else:
    print("Fail")
#正确的用户名,正确的密码
payload={"name":correctusername,"password":correctpassword}
data = requests.post(url,data=payload)
```

```
    if (str(data.status_code)=="200") and ("查看购物车" in str(data.text)):
        print("pass")
else:
    print("Fail")
```

这样的代码虽然可以测试,但是,没有测试框架的代码是不利于维护的,也不利于批量执行。为了解决这个问题,可以用刚才介绍的 unittest 框架进行改造。

testLogin.py

```
#!/usr/bin/env python
#coding:utf-8
import unittest,requests

class CheckUserUnit(unittest.TestCase):

    def setUp(self):
        self.correctusername="jerry"
        self.correctpassword="123456"
        self.discorrectusername="tom"
        self.discorrectpassword="000000"
        self.url="http://localhost:8080/sec/20/jsp/index.jsp"

    #错误的用户名,正确的密码
    def test_login_eucp(self):
        payload={"name": self.discorrectusername,"password": self.correctpassword}
        data = requests.post(self.url,data=payload)
        self.assertEqual("200",str(data.status_code))
        self.assertIn("用户名或者密码错误",str(data.text))

    #正确的用户名,错误的密码
    def test_login_cuep(self):
        payload={"name": self.correctusername,"password": self.discorrectpassword}
        data = requests.post(self.url,data=payload)
        self.assertEqual("200",str(data.status_code))
        self.assertIn("用户名或者密码错误",str(data.text))

    #错误的用户名,错误的密码
    def test_login_euep(self):
        payload={"name":self.discorrectusername,"password":self.discorrectpassword}
        data = requests.post(self.url,data=payload)
        self.assertEqual("200",str(data.status_code))
        self.assertIn("用户名和密码错误",str(data.text))

    #正确的用户名,正确的密码
    def test_login_cucp(self):
```

```
            payload={"name": self.correctusername,"password":self.correctpassword}
            data = requests.post(self.url,data=payload)
            self.assertEqual("200",str(data.status_code))
            self.assertIn("查看购物车",str(data.text))


if __name__=='__main__':
    #构造测试集
    suite=unittest.TestSuite()
    suite.addTest(CheckUserUnit("test_login_eucp"))
    suite.addTest(CheckUserUnit("test_login_cuep"))
    suite.addTest(CheckUserUnit("test_login_euep"))
    suite.addTest(CheckUserUnit("test_login_cucp"))
    #运行测试集合
    runner=unittest.TextTestRunner()
    runner.run(suite)
```

程序通过 self.assertEqual("200",str(data.status_code))判断返回码是不是与预期的相同；通过 self.assertIn("用户名和密码错误",str(data.text))判断返回的文本中是不是包括指定的字符串。测试用例 test_login_eucp、test_login_cuep 和 test_login_euep 为错误情况的测试用例，将在返回页面中出现"用户名或者密码错误"的提示（注意，这里是返回的 HTTP 代码中的字符串，而不是页面显示字符串），test_login_cucp 为正确的测试用例，登录的用户名和密码都正确，系统跳到商品列表页面，并且显示"查看购物车"字符串。所以，在代码中以返回页面中是否存在"查看购物车"字符串判断测试是否成功。

2.11.4 数据驱动的自动化接口测试

数据驱动的自动化接口测试是 HP 公司在其著名的产品 QTP 中提出的，并且成为业内自动化测试的一个标准。数据驱动可以理解为测试数据的参数化。由于 Python 读取 XML 的技术相当成熟，所以可以把测试数据放在 XML 里，然后进行设计数据驱动的自动化接口测试（当然，数据也可以放在文本文件、JSON 文件或数据库中）。下面介绍如何设计 XML 文件。

loginConfig.xml：

```
<?xml version="1.0" encoding="UTF-8"?>
<node>
    <case>
        <TestId>testcase001</TestId>
        <Title>用户登录</Title>
        <Method>post</Method>
        <Desc>正确用户名,错误密码</Desc>
        <Url>http://localhost:8080/sec/20/jsp/index.jsp</Url>
        <InptArg>{"name":"jerry","password":"000000"}</InptArg>
```

```xml
        <Result>200</Result>
        <CheckWord>用户名或者密码错误</CheckWord>
    </case>
    <case>
        <TestId>testcase002</TestId>
        <Title>用户登录</Title>
        <Method>post</Method>
        <Desc>错误用户名,正确密码</Desc>
        <Url>http://localhost:8080/sec/20/jsp/index.jsp</Url>
        <InptArg>{"name":"smith","password":"123456"}</InptArg>
        <Result>200</Result>
        <CheckWord>用户名或者密码错误</CheckWord>
    </case>
    <case>
        <TestId>testcase003</TestId>
        <Title>用户登录</Title>
        <Method>post</Method>
        <Desc>错误用户名,错误密码</Desc>
        <Url>http://localhost:8080/sec/20/jsp/index.jsp</Url>
        <InptArg>{"name":"smith","password":"000000"}</InptArg>
        <Result>200</Result>
        <CheckWord>用户名和密码错误</CheckWord>
    </case>
    <case>
        <TestId>testcase004</TestId>
        <Title>用户登录</Title>
        <Method>post</Method>
        <Desc>正确用户名,正确密码</Desc>
        <Url>http://localhost:8080/sec/20/jsp/index.jsp</Url>
        <InptArg>{"name":"jerry","password":"123456"}</InptArg>
        <Result>200</Result>
        <CheckWord>查看购物车</CheckWord>
    </case>
</node>
```

这里,<node>…</node>是根标识,<case>…</case>表示一个测试用例,这里面有四个<case>…</case>对,分别表示上述四个测试用例。在<case>…</case>对中,有些数据是为了代码阅读者阅读起来更加方便,有些数据是程序中要是用的,下面对它们进行介绍。

(1) <TestId>…</TestId>:标号。

(2) <Title>…</Title>:标题(便于阅读)。

(3) <Method>…</Method>:传输方法,post/get。

(4) <Desc>…</Desc>:测试用例描述(便于阅读)。

(5) <Url>…</Url>:测试的 URL 地址。

(6) <InptArg>…</InptArg>：请求参数，用{}括起来，为符合 Python 字典格式的值参对。

(7) <Result>…</Result>：返回码。

(8) <CheckWord>…</CheckWord>：验证字符串。

首先介绍 Python 是如何获得 XML 中的内容的。假设有一个 XML 参数对<AAA>…</AAA>，在 .py 文件中

(1) 通过调用 from xml.dom import minidom 引入 minidom 类。

(2) 通过 dom=minidom.parse('loginConfig.xml') 获取需要读取的 xml 文件。

(3) 通过 root = dom.documentElement 开始获取文件中节点的内容。

(4) 然后通过语句 aaa = root.getElementsByTagName('AAA') 获得文件中的所有叶子节点<AAA>…</AAA>对中的数据。

(5) 但是，由于 XML 文件中的标签往往不止一个，且成对出现，正像文件 loginConfig.xml 中的<TestId>…</TestId>、<Title>…</Title> 、<Method>…</Method>…，所以可以通过 Python 的 for 循环语句获得，如下列代码所示。

```python
aaa=root.getElementsByTagName('AAA')
bbb=root.getElementsByTagName('BBB')
ccc=root.getElementsByTagName('CCC')
i=0
for key in AAA:
    aaaValue =(aaa[i].firstChild.data).strip()
    bbbValue =(bbb[i].firstChild.data).strip()
    cccValue =(ccc[i].firstChild.data).strip()
    print(aaaValue)
    print(bbbValue)
    print(cccValue)
    i = i+1
```

接下来介绍测试代码。

loginConfig.py

```python
#!/usr/bin/env python
#coding:utf-8
import unittest,requests
from xml.dom import minidom

class mylogin(unittest.TestCase):
    def setUp(self):
        print("--------测试开始--------")
        #从 XML 中读取数据
        dom =minidom.parse('loginConfig.xml')
        root =dom.documentElement
```

```python
            TestIds = root.getElementsByTagName('TestId')
            Titles = root.getElementsByTagName('Title')
            Methods = root.getElementsByTagName('Method')
            Descs = root.getElementsByTagName('Desc')
            Urls = root.getElementsByTagName('Url')
            InptArgs = root.getElementsByTagName('InptArg')
            Results = root.getElementsByTagName('Result')
            CheckWords = root.getElementsByTagName('CheckWord')
            i = 0
            mylists= []
            for TestId in TestIds:
                    mydicts={}
                    #获取每个数据,形成字典
                    mydicts["TestId"] = (TestIds[i].firstChild.data).strip()
                    mydicts["Title"] = (Titles[i].firstChild.data).strip()
                    mydicts["Method"] = (Methods[i].firstChild.data).strip()
                    mydicts["Desc"] = (Descs[i].firstChild.data).strip()
                    mydicts["Url"] = (Urls[i].firstChild.data).strip()
                    if ((InptArgs[i].firstChild) is None):
                            mydicts["InptArg"] =""
                    else:
                            mydicts["InptArg"]= (InptArgs[i].firstChild.data).strip()
                    mydicts["Result"] = (Results[i].firstChild.data).strip()
                    mydicts["CheckWord"]= (CheckWords[i].firstChild.data).strip()
                    mylists.append(mydicts)
                    i = i+1
            self.mylists =mylists
    def test_login(self):
            for mylist in self.mylists:
                    payload = eval(mylist["InptArg"])
                    url=mylist["Url"]
                    #发送请求
                    try:
                            if mylist["Method"] =="post":
                                    data = requests.post(url,data=payload)
                            elif mylist["Method"] =="get":
                                    data = requests.get(url,params=payload)
                            else:
                                    print ("Method参数获取错误")
                    except Exception as e:
                            self.assertEqual(mylist["Result"],"404")
                    else:
                            self.assertEqual(mylist["Result"],str(data.status_code))
                            self.assertIn(mylist["CheckWord"],str(data.text))
```

```
            def tearDown(self):
                    print("--------测试结束--------")
if __name__=='__main__':
    #构造测试集
    suite=unittest.TestSuite()
    suite.addTest(mylogin("test_login"))
    #运行测试集合
    runner=unittest.TextTestRunner()
    runner.run(suite)
```

由于标签 InptArg 里面可以没有参数,所以这里用 if ((InptArgs[i].firstChild) is None)进行判断,如果是 None,就把变量 mydicts["InptArg"]赋为空串。

setUp(self)主要把 XML 里的所有叶子节点数据获取到,放在一个名为 mylists 的列表变量中,最后返回给 self.mylists 变量。列表中的每一项为一个字典类型的数据,key 为 XML 里的所有叶子节点标签,key 所对应的值为 XML 标签中的内容。最后,self.mylists 传到每个测试方法中使用。

现在来看方法 test_login(self)。

(1) for mylist in self.mylists:把刚才在初始化里定义的 self.mylists 的每一项分别获取出来。

(2) payload = eval(mylist["InptArg"]):获取标签为 InptArg 中的数据,由于在 XML 格式定义的时候,这一项用{}括起来,里面是一个值参对,又由于 mylist["InptArg"]返回的是一个具有字典格式的变量,所以必须通过方法 eval()转义成字典变量赋给 payload。

(3) url=mylist["Url"]为接口测试请求 HTTP 的 URL 地址。

(4) 通过判断 mylist["Method"]等于 post,还是等于 get,选择使用 data = requests.post(url,data=payload)或者 data = requests.get(url,params=payload)发送信息,接受信息放在变量 data 中。

(5) 通过 self.assertEqual(mylist["Result"],str(data.status_code))判断返回码是否符合期望结果,以及通过 self.assertIn(mylist["CheckWord"],str(data.text))判断期望的字符串 mylist["CheckWord"]是否在返回的内容 str(data.text)中存在,从而判断测试是否成功。这里特别指出,在程序 except Exception as e 中通过 self.assertEqual(mylist["Result"],"404")判断期望结果的返回状态码是否为 404。在这个项目中加上 runtest.py 运行所有的测试用例。格式与前面相同,这里不再重复介绍。图 2-17 是基于 Python Requests 的 HTTP 接口测试报告。

2.11.5 进一步优化

细心的读者会发现,上面程序中的 setUp()方法可以进行进一步的封装优化,建立一个单独的.py 文件 getXML.py,具体内容如下。

测试报告

Start Time: 2017-08-29 11:44:36
Duration: 0:00:00.230844
Status: Pass 2

测试用例执行报告

Show Summary Failed All

Test Group/Test case	Count	Pass	Fail	Error	View
loginTest.mylogin	1	1	0	0	Detail
test_login			pass		
registerTest.mylogin	1	1	0	0	Detail
test_login			pass		
Total	2	2	0	0	

图 2-17 基于 Python Requests 的 HTTP 接口测试报告

```python
#!/usr/bin/env python
#coding:utf-8
from xml.dom import minidom

class GetXML:
    def getxmldata(xmlfile):
        #从 XML 中读取数据
        dom = minidom.parse(xmlfile)
        root = dom.documentElement
        TestIds = root.getElementsByTagName('TestId')
        Titles = root.getElementsByTagName('Title')
        Methods = root.getElementsByTagName('Method')
        Descs = root.getElementsByTagName('Desc')
        Urls = root.getElementsByTagName('Url')
        InptArgs = root.getElementsByTagName('InptArg')
        Results = root.getElementsByTagName('Result')
        CheckWords = root.getElementsByTagName('CheckWord')
        i = 0
        mylists=[]
        for TestId in TestIds:
            mydicts={}
            #获取每个数据,形成字典
            mydicts["TestId"] = (TestIds[i].firstChild.data).strip()
            mydicts["Title"] = (Titles[i].firstChild.data).strip()
            mydicts["Method"] = (Methods[i].firstChild.data).strip()
            mydicts["Desc"] = (Descs[i].firstChild.data).strip()
            mydicts["Url"] = (Urls[i].firstChild.data).strip()
            if ((InptArgs[i].firstChild) is None):
                mydicts["InptArg"] = ""
            else:
                mydicts["InptArg"] = (InptArgs[i].firstChild.data).strip()
            mydicts["Result"] = (Results[i].firstChild.data).strip()
```

```
                mydicts["CheckWord"] = (CheckWords[i].firstChild.data).strip()
                mylists.append(mydicts)
                i = i+1
        return mylists
```

这样，在 LoginTest.py 中的 setUp()方法只需要进行如下修改就可以了。

```
...
from getXML import GetXML  #引入刚才建立的类
...
class mylogin(unittest.TestCase):
        def setUp(self):
                print("--------测试开始--------")
                self.mylists = GetXML.getxmldata("loginConfig.xml")  #调用类中的方法
...
```

第3章 电子商务网站的实现

3.1 需求描述

3.1.1 用户信息模块

用户信息模块包括"用户信息的注册""用户登录""显示用户信息"和"用户密码的修改"。

(1) 注册信息要输入用户名、密码和邮箱。注册信息要求用户名必须唯一,如果用户名在数据库中已经存在,就会显示相应的错误提示信息。

(2) 在用户登录的时候,如果用户名和密码输入有误,就必须提示相应的错误信息。

(3) 用户登录程序后,应该允许用户查看自己的用户信息和收货信息。

(4) 允许修改密码,修改用户密码的时候,必须提供旧密码、新密码和新密码的确认信息。下列情况应该给出相应的错误提示信息。

① 旧密码不正确。

② 新密码与旧密码相同。

③ 新密码与新密码的确认信息不一致。

3.1.2 商品信息模块

商品信息模块包括"商品信息的维护""商品概要信息的分页显示""根据商品名称的模糊查询"和"对某一条商品显示其详细信息"。

(1) "商品信息的维护":包括增加、修改和删除操作,是利用 Django 的后台完成的。

(2) "商品概要信息的分页显示":包括显示商品信息的 id、名称、价钱以及查看详情和放入购物车的操作链接。

(3) "根据商品名称的模糊查询":通过商品名称的模糊查询实现,查询结果界面同概要信息,也需要实现分页功能。

(4) "对某一条商品显示其详细信息":除了显示名称、价钱,还要显示商品的描述、图片以及放入购物车的操作。

3.1.3 购物车模块

购物车模块包括"购物车中所有商品的显示""添加商品进入购物车""修改购物车中某

种商品的数量""删除购物车中某个商品"和"删除购物车中所有商品"。

（1）"购物车中所有商品的显示"通过列表实现，包括显示商品 id、商品名称、单价、商品个数以及移除的操作链接。单击"商品 id"可以查看对应的商品详细信息。

（2）"添加商品进入购物车"可以在购物车列表中进行操作，也可以在商品的详细信息中进行操作。

（3）"修改购物车中某种商品的数量"和"删除购物车中某个商品"的操作在购物车列表中进行。

（4）可以在查看所有订单页面中"删除购物车中所有商品"。

3.1.4　送货地址模块

送货地址模块包括"送货地址的显示""送货地址的添加""送货地址的修改"和"送货地址的删除"。

（1）"送货地址的显示"可以在生成订单选择送货地址的时候，也可以在查看用户信息的时候。

（2）"送货地址的添加"可以添加当前用户账号下的一个或多个送货地址。

（3）"送货地址的修改"和"送货地址的删除"可以通过送货地址的显示页面进入。

3.1.5　订单模块

订单模块包括"显示总的订单""显示所有订单""删除单个订单"以及"删除总订单"。

（1）"显示总的订单"在订单生成完毕后显示，包括生成时间、配货地址和总价钱以及订单中每个商品的订单 id、商品名称、商品价格、个数。

（2）"显示所有订单"包括该用户下的所有订单，每个订单的显示内容同单个订单。如果这个订单没有支付，系统就会提供支付的操作链接。

（3）"删除单个订单"可以在显示单个订单内容页面中进行，也可以在显示所有订单内容页面中进行。

（4）"删除总订单"在显示单个订单或显示所有订单的页面中进行。

（5）在单个订单和所有订单中单击"商品 id"可以查看对应的商品详细信息。

3.1.6　订单支付模块

订单确认后，可以利用各种支付平台（如支付宝、微信、网银卡等）进行支付操作。

3.2　数据 Model 设计

根据 3.1 节的需求描述进行数据模型设计。电子商务系统的系统关联图（E-R 图）如图 3-1 所示。

这里建立了 5 个对象，分别是用户（User）、地址（Address）、商品（Goods）、单个订单（Order）和总订单（Orders）。

（1）一个用户对应多个地址，一个地址对应一个用户，所以"用户，地址"是一对多的关系，需要在地址表中建立包含指向用户表的外键。

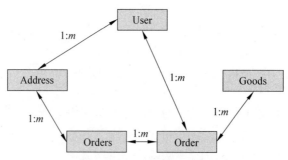

图 3-1　电子商务系统的系统关联图(E-R 图)

(2) 一个地址对应多个总订单,一个总订单对应一个地址,所以"地址,总订单"是一对多的关系,需要在总订单表中建立包含指向地址表的外键。

(3) 一个用户对应多个订单,一个订单对应一个用户,所以"用户,订单"是一对多的关系,需要在订单表中建立包含指向用户表的外键。

(4) 一个商品对应多个单个订单,一个单个订单对应一个商品,所以"商品,单个订单"是一对多的关系,需要在单个订单表中建立包含指向商品的外键。

(5) 一个总订单对应多个单个订单,一个单个订单对应一个总订单,所以"总订单,单个订单"是一对多的关系,需要在单个订单表中建立包含指向总订单的外键。

根据上述分析,建立如下 model.py 文件。

```python
from django.db import models

#在这里建立模型
#用户
class User(models.Model):
    username = models.CharField(max_length=50)          #用户名
    password = models.CharField(max_length=50)          #密码
    email = models.EmailField()                         #Email

    def __str__(self):
        return self.username

#商品
class Goods(models.Model):
    name = models.CharField(max_length=100)             #商品名称
    price = models.FloatField()                         #单价
    picture = models.FileField(upload_to = './upload/') #图片
    desc = models.TextField()                           #描述

    def __str__(self):
        return self.name
```

```python
#收货地址
class Address(models.Model):
    user = models.ForeignKey(User)                          #关联用户id
    address = models.CharField(max_length=50)               #地址
    phone = models.CharField(max_length=15)                 #电话

    def __str__(self):
        return self.address

#总订单
class Orders(models.Model):
    address = models.ForeignKey(Address)                    #关联送货地址id
    create_time = models.DateTimeField(auto_now=True)       #创建时间
    status = models.BooleanField()                          #订单状态

    def __str__(self):
        return self.create_time

#单个订单
class Order(models.Model):
    order = models.ForeignKey(Orders)                       #关联总订单id
    user = models.ForeignKey(User)                          #关联用户id
    goods = models.ForeignKey(Goods)                        #关联商品id
    count = models.IntegerField()                           #数量
```

Goods 表中 picture 使用的是 models.FileField(upload_to = './upload/')，也可以用 ImageField，由于 goods 是后台管理人员操作的，为了提高上传图片的性能，没有使用 ImageField 管理。upload_to = './upload/'表示图片上传后，放入名为 upload 的路径。upload 路径是与 goods 平行的。这样就需要在 urls.py 中加入如下代码。

```
...
url(r'^static/(?P<path>.*)',static.serve,{'document_root':os.path.join(BASE_
DIR,'upload')}),
...
```

通过后台上传的图片文件自动存在 BASE_DIR\upload\下，通过＜image src='/upload/…jpg'＞显示相应的图片(再次提醒，upload 前必须有字符'/')。

3.3 用户信息模块

用户信息模块包括"用户信息的注册""用户登录""显示用户信息"和"用户密码的修改"。其中，"用户信息的注册"与"用户登录"在本书第 2 章已进行了详细描述，本章将对它们进行系统的归纳与优化。数据模型如下。

```
...
#用户
class User(models.Model):
    username = models.CharField(max_length=50)       #用户名
    password = models.CharField(max_length=50)       #密码
    email = models.EmailField()                      #Email

    def __str__(self):
        return self.username
...
```

3.3.1 用户注册

只有通过用户注册的用户才可以登录系统，根据需求，在这个系统中用户注册需要填写用户名、密码和 Email 地址。

1. urls.py

```
...
url(r'^register/$', views.register),
...
```

2. views.py

把所有的表单定义在一个名为 forms.py 的文件中，用户注册的表单定义如下。

```
...
from django import forms

#定义注册表单模型
class UserForm(forms.Form):
    username = forms.CharField(label='用户名', max_length=100)
    password = forms.CharField(label='密码', widget=forms.PasswordInput())
    email = forms.EmailField(label='电子邮件')
...
```

这里，

(1) username 中的 max_length=100 表示输入的字符个数最多为 100 个。

(2) password 中的 widget=forms.PasswordInput() 表示文本信息为密码格式。

(3) email 中的 EmailField 表示格式为 HTML5 中的 Email 格式。

然后在 views.py 中通过 from goods.forms import UserForm 引入。下面是 views.py 中关于注册的代码。

```
...
from goods.forms import UserForm
...
```

```python
#用户注册
def register(request):
    if request.method == "POST":                      #判断表单是否提交状态
        uf = UserForm(request.POST)                   #获得表单变量
        if uf.is_valid():                             #判断表单数据是否正确
            #获取表单信息
            username = (request.POST.get('username')).strip()   #获取用户名信息
            password = (request.POST.get('password')).strip()   #获取密码信息
            email = (request.POST.get('email')).strip()         #获取 Email 信息
            #查找数据库中是否存在相同的用户名
            user_list = User.objects.filter(username=username)
            if user_list:
                #如果存在,就报"用户名已经存在!"错误信息,并且回到注册页面
                return render_to_response('register.html',{'uf':uf,"error":"用户名已经存在!"})
            else:
                #否则将表单写入数据库
                user = User()
                user.username = username
                user.password = password
                user.email = email
                user.save()
                #返回登录页面
                uf = LoginForm()
                return render_to_response('index.html',{'uf':uf})
    else:   #如果不是表单提交状态,就显示表单信息
        uf = UserForm()
    return render_to_response('register.html',{'uf':uf})
...
```

(1) 通过 if request.method == "POST":判断当前状态是否为表单提交状态,如果不是,就显示表单注册页面:uf = UserForm(),return render_to_response('register.html',{'uf':uf}),否则验证提交的表单信息:if uf.is_valid():。

(2) 判断表单提交是否正确,如果正确,就获取提交的信息:username = (request.POST.get('username')).strip()和 password = (request.POST.get('password')).strip()。

(3) 通过 user_list = User.objects.filter(username=username)的返回变量 user_list 是否为空判断注册的用户名是否已经被注册过,如果未注册过,就提示错误信息,否则接受提交的注册信息,通过 user.save()将其保存在数据库中。

这里特别需要注意的是,由于后面需要用到基于 Requests 的接口测试,这里必须使用 request.POST.get('username')获取表单数据。

3. 模板

模板文件为 register.html,其内容如下。

```
{%load staticfiles%}
<!DOCTYPE html>
<html lang="zh-CN">
    <head>
        <meta charset="utf-8">
        <meta http-equiv="X-UA-Compatible" content="IE=edge">
        <meta name="viewport" content="width=device-width, initial-scale=1">
        <!--上述 3 个 meta 标签 * 必须 * 放在最前面,任何其他内容都 * 必须 * 跟随其后!-->
        <meta name="description" content="">
        <meta name="author" content="">

        <title>电子商务系统-注册</title>

        <!--Bootstrap core CSS -->
        <link href="{%static 'css/signin.css'%}" rel="stylesheet"-->
        <!--Custom styles for this template -->
        <link href="{%static 'css/bootstrap.min.css'%}" rel="stylesheet"-->
        <link href="{%static 'css/my.css'%}" rel="stylesheet">

    </head>

    <body>

        <div class="container">
            <form class="form-signin" method="post" enctype="multipart/form-data">
                <h2 class="form-signin-heading">电子商务系统-注册</h2>
                {{uf.as_p}}
                    <p style="color:red">{{error}}</p><br>
                <button class="btn btn-lg btn-primary btn-block" type="submit">注册</button><br>
                <a href="/index/">登录</a>
            </form>

        </div><!--/container -->

    </body>
</html>
```

其中,

(1){{error}}:显示的是错误提示信息。

(2){{uf.as_p}}:显示的是表单信息。

注册页面如图 3-2 所示。

4. 接口测试

这里在本书 2.11 节的基础上对测试方法进行了一些优化,主要通过利用 Python 对数

图 3-2 注册页面

据库的访问与接口测试相结合的方法进行相应的测试。

1) 测试用例

表 3-1 为注册模块的测试用例,这里共设计了两个用例。

(1) 注册一个数据库中已经存在的用户,系统应该提示"用户名已经存在!"。

(2) 注册一个数据库中不存在的用户,注册成功,进入登录页面。

表 3-1 注册模块的测试用例

编号	描述	期望结果
1	注册的用户名已经存在	有提示信息"用户名已经存在!"
2	注册的用户名不存在	注册成功,进入登录页面

2) XML 数据文件

loginRegConfig.xml

```xml
<node>
    <case>
        <username>Johnson</username>
        <password>12345</password>
        <email>Johnson@126.com</eamil>
    </case>
    <case>
        <TestId>loginReg-testcase001</TestId>
        <Title>用户注册</Title>
        <Method>post</Method>
        <Desc>注册用户名已经存在</Desc>
        <Url>http://127.0.0.1:8000/register/</Url>
        <InptArg>{"username":"Johnson","password":"12345","email":"Johnson5@126.com"}</InptArg>
        <Result>200</Result>
        <CheckWord>用户名已经存在!</CheckWord><!--通过"注册用户名不存在"测试完毕
删除数据库记录 -->
```

```xml
        </case>
        <case>
            <TestId>loginReg-testcase002</TestId>
            <Title>用户注册</Title>
            <Method>post</Method>
            <Desc>注册用户名不存在</Desc>
            <Url>http://127.0.0.1:8000/register/</Url>
            <InptArg>{"username":"smith","password":"12345","email":"smith@126.com"}</InptArg>
            <Result>200</Result>
            <CheckWord>登录</CheckWord>
        </case>
</node>
```

第一个＜case＞…＜/case＞为测试代码用的初始化信息,将通过测试程序中的 setUp() 中 Python 语言的基础类 sqlite3(注意,这里不是通过 Django 提供的数据库操作模块)向数据库中插入记录,然后运行程序进行测试,最后测试结束,需要在 tearDown() 方法中将这些记录删除。在后面所有模块测试代码中都采用这种方法。

3) 测试代码

为了方便,首先对一些方法进行封装。把本书 2.11.5 节中 getXML.py 中的类 GetXML 封装在一个名为 util.py 的文件中,并且把头部的两行建立在这个类的构造方法中。

```python
def __init__(self,myXmlFile):
        dom = minidom.parse(myXmlFile)
        self.root = dom.documentElement
```

这样,原先的 getxmldata() 方法就改为

```python
def getxmldata(self):
        #从 XML 中读取数据
        TestIds = self.root.getElementsByTagName('TestId')
        Titles = self.root.getElementsByTagName('Title')
        Methods = self.root.getElementsByTagName('Method')
        Descs = self.root.getElementsByTagName('Desc')
        Urls = self.root.getElementsByTagName('Url')
        InptArgs = self.root.getElementsByTagName('InptArg')
        Results = self.root.getElementsByTagName('Result')
        CheckWords = self.root.getElementsByTagName('CheckWord')
        i = 0
        mylists=[]
        for TestId in TestIds:
                mydicts={}
                #获取每个数据,形成字典
```

```
                mydicts["TestId"] = (TestIds[i].firstChild.data).strip()
                mydicts["Title"] = (Titles[i].firstChild.data).strip()
                mydicts["Method"] = (Methods[i].firstChild.data).strip()
                mydicts["Desc"] = (Descs[i].firstChild.data).strip()
                mydicts["Url"] = (Urls[i].firstChild.data).strip()
                mydicts["InptArg"] = (InptArgs[i].firstChild.data).strip()
                mydicts["Result"] = (Results[i].firstChild.data).strip()
                mydicts["CheckWord"]= (CheckWords[i].firstChild.data).strip()
                mylists.append(mydicts)
                i = i+1
        return mylists
```

然后在这个类中获得 User 测试初始化信息的方法 getUserInitInfo(),具体实现如下。

```
def getUserInitInfo(self):
        #从 XML 中读取数据
        id = self.root.getElementsByTagName('id')
        id = (str(id[0].firstChild.data)).strip()
        username = self.root.getElementsByTagName('username')
        username = "\""+(str(username[0].firstChild.data)).strip()+"\""
        password = self.root.getElementsByTagName('password')
        password = "\""+(str(password[0].firstChild.data)).strip()+"\""
        email = self.root.getElementsByTagName('email')
        email = "\""+(str(email[0].firstChild.data)).strip()+"\""
        values = id +","+username+","+password+","+email
        return values     #返回的字符串 values 供插入数据库表 good_user 中使用
```

最后形成的字符串 values 为插入 User 表中 SQL 语句 values 后的内容。然后在这个文件中建立一个 DB 类,主要用于封装实现对数据库的操作,现在先建立以下几个方法。

```
class DB:
        #构造方法,获得 sqlite3 数据库文件的位置
        def __init__(self):
                self.url ="C:\\Python35\\Scripts\\ebusiness\\db.sqlite3"

        #连接数据库连接
        def connect(self):
                self.con = con = sqlite3.connect(self.url)
                self.cur = self.con.cursor()

        #关闭数据库连接
        def close(self):
                self.cur.close()
                self.con.close()
```

```python
#向 tablename 表中插入数据 values
def insert(self,tablename,values):
    sql ="insert into "+tablename+" values ("+values+")"
    self.con.execute(sql)
    self.con.commit()

#在 tablename 表中删除满足 condition 条件的记录
def delete(self,tablename,condition):
    sql ="delete from "+tablename+" where ("+condition+")"
    self.con.execute(sql)
    self.con.commit()
```

(1) 方法 init()用于初始化数据库。
(2) 方法 connect()用于连接数据库。
(3) 方法 close()用于关闭数据库的连接。
(4) 方法 insert()用于向数据库表中插入数据。
(5) 方法 delete()用于从数据库表中删除满足条件的数据。

最后介绍用户注册模块的测试代码。

```python
#!/usr/bin/env python
#coding:utf-8
import unittest,requests
from util import GetXML,DB

class myregister(unittest.TestCase):

    def setUp(self):
        print("--------测试开始--------")
        #定义数据库表名
        self.userTable ="goods_user"
        #建立 GetXML 对象变量
        xmlInfo =GetXML("registerConfig.xml")
        #获得初始化信息
        self.userValues =xmlInfo.getUserInitInfo()
        #建立 DB 对象变量
        self.dataBase=DB()
        #连接数据库
        self.dataBase.connect()
        #插入初始化数据库
        self.dataBase.insert(self.userTable,self.userValues)
        #获得所有测试数据
        self.mylists =xmlInfo.getxmldata()

    def test_register(self):
```

```python
            for mylist in self.mylists:
                #获取传输参数
                payload=eval(mylist["InptArg"])
                #获取测试URL
                url=mylist["Url"]
                #发送请求
                try:
                    if mylist["Method"]=="post":
                        data=requests.post(url,data=payload)
                    elif mylist["Method"]=="get":
                        data=requests.get(url,params=payload)
                    else:
                        print("Method参数获取错误")
                except Exception as e:
                    self.assertEqual(mylist["Result"],"404")
                else:
                    #验证返回码
                    self.assertEqual(mylist["Result"],str(data.status_code))
                    #验证返回文本
                    self.assertIn(mylist["CheckWord"],str(data.text))

    def tearDown(self):
        #获取初始化数据库中的记录主码
        id=self.userValues.split(',')[0]
        #删除这条记录
        self.dataBase.delete(self.userTable,"id="+id)
        #关闭数据库连接
        self.dataBase.close()
        print("--------测试结束--------")


if __name__=='__main__':
    #构造测试集
    suite=unittest.TestSuite()
    suite.addTest(myregister("test_register"))
    #运行测试集合
    runner=unittest.TextTestRunner()
    runner.run(suite)
```

(1) 在 setUp 方法中,

① 先定义在这里用到的数据库表名为 goods_user。

② 然后通过语句 xmlInfo = GetXML("registerConfig.xml")从 XML 文件中读入测试初始化数据,并且通过 self.userValues = xmlInfo.getUserInitInfo()把它放到变量 self.userValues 中。

③ 之后建立数据库连接,通过语句 self.dataBase.insert(self.userTable,self.userValues)向数据库中插入设置测试需要的初始化信息。

④ 最后通过语句 self.mylists = xmlInfo.getxmldata()获得测试数据。

(2) 在测试程序中,

① 通过循环语句 for mylist in self.mylists 遍历所有的测试数据。

② 通过语句 payload = eval(mylist["InptArg"])获取测试参数,以及通过语句 url=mylist["Url"]获取测试执行路径。

③ 通过判断语句 mylist["Method"]是 post,还是 get 调用 data = requests.post(url,data=payload)或者 data=requests.get(url,params=payload)。

④ 通过断言语句 self.assertEqual(mylist["Result"],str(data.status_code))验证返回码是否正确。

⑤ 通过断言语句 self.assertIn(mylist["CheckWord"],str(data.text))判断验证的字符串是否在返回的文本中。

(3) 在 tearDown 方法中,

① 通过语句 id = self.userValues.split(',')[0]初始化数据库用户数据的主键。

② 通过语句 self.dataBase.delete(self.userTable,"id="+id)删除这条测试数据。

③ 断开数据库连接,结束测试。

3.3.2 用户登录

注册的用户可以通过登录页面登录系统。由于这个模块在前面讲得比较多,这里不做过多的解释。

1. urls.py

```
...
url(r'^$', views.index),
url(r'^index/$', views.index),
url(r'^ login_action /$', views.login_action),
...
```

2.6.3 节提到登录页面为系统首页,提供了 3 个 URL,分别对应

(1) 127.0.0.1:8000/。

(2) 127.0.0.1:8000/index/。

(3) 127.0.0.1:8000/login_action/。

2. views.py

```
...
#首页(登录)
def index(request):
    uf = LoginForm()
    return render_to_response('index.html',{'uf':uf})
...
#用户登录
```

```python
def login_action(request):
    if request.method == "POST":
        uf = LoginForm(request.POST)
        if uf.is_valid():
            #寻找名为 "username"和"password"的 POST 参数,而且如果参数没有提交,就返回一个空的字符串
            username = (request.POST.get('username')).strip()
            password = (request.POST.get('password')).strip()
            #判断输入数据是否为空
            if username =='' or password =='':
                return render(request,"index.html",{'uf':uf,"error":"用户名和密码不能为空"})
            else:
                #判断用户名和密码是否准确
                user = User.objects.filter(username=username,password=password)
                if user:
                    response = HttpResponseRedirect('/goods_view/')
                    #登录成功后跳转查看商品信息
                    request.session['username'] = username  #将 session 信息写到服务器
                    return response
                else:
                    return render(request,"index.html",{'uf':uf,"error":"用户名或者密码错误"})
        else:
            uf = LoginForm()
    return render_to_response('index.html',{'uf':uf})
...
```

3. 模板

模板文件为 index.html,其内容如下。

```
{%load staticfiles%}
<!DOCTYPE html>
<html lang="zh-CN">
    <head>
        <meta charset="utf-8">
        <meta http-equiv="X-UA-Compatible" content="IE=edge">
        <meta name="viewport" content="width=device-width, initial-scale=1">
        <!--上述 3 个 meta 标签 * 必须 * 放在最前面,任何其他内容都 * 必须 * 跟随其后!-->
        <meta name="description" content="">
        <meta name="author" content="">
        <link rel="icon" href="../../favicon.ico">

        <title>电子商务系统-登录</title>
```

```html
        <!--Bootstrap core CSS -->
        <link href="{%static 'css/signin.css'%}" rel="stylesheet"-->
        <!--Custom styles for this template -->
        <link href="{%static 'css/bootstrap.min.css'%}" rel="stylesheet"-->
        <link href="{%static 'css/my.css'%}" rel="stylesheet">

    </head>

    <body>

        <div class="container">
            <form class="form-signin" method="post" action="/login_action/" enctype="multipart/form-data">
                <h2 class="form-signin-heading">电子商务系统-登录</h2>
                {{uf.as_p}}
                <p style="color:red">{{error}}</p><br>
                <button class="btn btn-lg btn-primary btn-block" type="submit">登录</button><br>
                <a href="\register\">注册</a>
            </form>

        </div><!--/container -->

    </body>
```

(1) {{error}}：显示错误提示信息。

(2) {{uf.as_p}}：显示表单信息。

用户登录界面如图 3-3 所示。

图 3-3　用户登录界面

4. 接口测试

重新构造初始化数据,代码如下。

loginConfig.xml：

```
<node>
    <case>
```

```xml
            <id>0</id>
            <username>Johnson</username>
            <password>000000</password>
            <email>Johnson@126.com</email>
        </case>
        <case>
            <TestId>loginReg-testcase001</TestId>
            <Title>用户登录</Title>
            <Method>post</Method>
            <Desc>正确用户名,错误密码</Desc>
            <Url>http://127.0.0.1:8000/login_action/</Url>
            <InptArg>{"username":"Johnson","password":"123456"}</InptArg>
            <Result>200</Result>
            <CheckWord>用户名或者密码错误</CheckWord>
        </case>
        <case>
            <TestId>loginReg-testcase002</TestId>
            <Title>用户登录</Title>
            <Method>post</Method>
            <Desc>错误用户名,正确密码</Desc>
            <Url>http://127.0.0.1:8000/login_action/</Url>
            <InptArg>{"username":"smith","password":"000000"}</InptArg>
            <Result>200</Result>
            <CheckWord>用户名或者密码错误</CheckWord>
        </case>
        <case>
            <TestId>loginReg-testcase003</TestId>
            <Title>用户登录</Title>
            <Method>post</Method>
            <Desc>错误用户名,错误密码</Desc>
            <Url>http://127.0.0.1:8000/login_action/</Url>
            <InptArg>{"username":"smith","password":"123456"}</InptArg>
            <Result>200</Result>
            <CheckWord>用户名和密码错误</CheckWord>
        </case>
        <case>
            <TestId>loginReg-testcase004</TestId>
            <Title>用户登录</Title>
            <Method>post</Method>
            <Desc>正确用户名,正确密码</Desc>
            <Url>http://127.0.0.1:8000/login_action/</Url>
            <InptArg>{"username":"Johnson","password":"000000"}</InptArg>
            <Result>200</Result>
            <CheckWord>查看购物车</CheckWord>
        </case>
</node>
```

测试代码如下。

loginTest.py

```python
#!/usr/bin/env python
#coding:utf-8
import unittest,requests
from util import GetXML,DB

class mylogin(unittest.TestCase):
    def setUp(self):
        print("--------测试开始--------")
        #定义数据库表名
        self.userTable ="goods_user"
        #建立 GetXML 对象变量
        xmlInfo =GetXML("loginConfig.xml")
        #获得初始化信息
        self.userValues =xmlInfo.getUserInitInfo()
        #建立 DB 对象变量
        self.dataBase=DB()
        #连接数据库
        self.dataBase.connect()
        #插入初始化数据库
        self.dataBase.insert(self.userTable,self.userValues)
        #获得所有测试数据
        self.mylists =xmlInfo.getxmldata()

    def test_login(self):
        for mylist in self.mylists:
            #获取传输参数
            payload =eval(mylist["InptArg"])
            #获取测试 URL
            url=mylist["Url"]
            #发送请求
            try:
                if mylist["Method"] =="post":
                    data =requests.post(url,data=payload)
                elif mylist["Method"] =="get":
                    data =requests.get(url,params=payload)
                else:
                    print ("Method 参数获取错误")
            except Exception as e:
                self.assertEqual(mylist["Result"],"404")
            else:
                #验证返回码
                self.assertEqual(mylist["Result"],str(data.status_code))
```

```
                        #验证返回文本
                        self.assertIn(mylist["CheckWord"],str(data.text))

    def tearDown(self):
        #获取初始化数据库中的记录主码
        id=self.userValues.split(',')[0]
        #删除这条记录
        self.dataBase.delete(self.userTable,"id="+id)
        #关闭数据库连接
        self.dataBase.close()
        print("--------测试结束--------")


if __name__=='__main__':
    #构造测试集
    suite=unittest.TestSuite()
    suite.addTest(mylogin("test_login"))
    #运行测试集合
    runner=unittest.TextTestRunner()
    runner.run(suite)
```

3.3.3 用户信息显示

登录的用户可以单击自己的用户名,显示自己的注册信息以及自己的所有收货地址信息。

1. urls.py

```
...
url(r'^user_info/$', views.user_info),
...
```

2. views.py

```
...
#获取用户信息
def user_info(request):
    #检查用户是否登录
    util=Util()
    username=util.check_user(request)
    #如果没有登录,就跳转到首页
    if username=="":
        uf=LoginForm()
        return render(request,"index.html",{'uf':uf,"error":"请登录后再进入!"})
    else:
```

```
        #count 为当前购物车中商品的数量
        count =util.cookies_count(request)
        #获取登录用户信息
        user_list =get_object_or_404(User,username=username)
        #获取登录用户收货地址的所有信息
        address_list =Address.objects.filter(user_id=user_list.id)
        returnrender(request,"view_user.html",{"user": username,"user_info":
user_list,"address":address_list,"count":count})
...
```

关于检查用户是否通过合法途径登录,本书2.6.2节中已经作过详细介绍。

(1) 当检查当前用户为合法用户后,通过语句 count = util.cookies_count(request)调用 util 类中的 cookies_count()方法,显示当前用户的购物车内有多少商品。

(2) 通过语句 user_list = get_object_or_404(User,username=username)获取当前登录用户的信息。

(3) 然后通过语句 address_list = Address.objects.filter(user_id=user_list.id)获取当前用户的所有地址信息。

(4) 最后通过语句 return render(request,"view_user.html",{"user":username,"user_info":user_list,"address":address_list,"count":count})返回给模板文件 view_user.html,其中传过去的变量包括

① user:当前登录用户名。

② user_info:当前登录用户名信息。

③ address:当前登录用户的所有地址信息。

④ count:当前购物车内商品的数量。

3. 模板

view_user.html:

```
{% extends "base.html" %}
{% block content %}
        <li class="active"><a href="/view_chart/">查看购物车<font color=
"#FF0000">{{ count }}</font></a></li>
        </ul>
        <ul class="nav navbar-nav navbar-right">
            <li><a href="/user_info/">{{user}}</a></li>
            <li><a href="/logout/">退出</a></li>
        </ul>
        </div><!--/.nav-collapse -->
    </div>
</nav>

<div class="container theme-showcase" role="main">

    <div class="page-header">
```

```html
            <div id="navbar" class="navbar-collapse collapse">
            </div><!--/.navbar-collapse -->
    </div>

    <div class="row">
        <div class="col-md-6">
            用户名：{{user_info.username}}<br>
            Email：{{user_info.email}}<br>
            <form action="/change_password/" method="get">
                <input type="submit" value="修改密码">
            </form>
            <table class="table table-striped">
                <thead>
                    <tr>
                        <th>编号</th>
                        <th>地址</th>
                        <th>电话</th>
                        <th>修改</th>
                        <th>删除</th>
                    </tr>
                </thead>
                <tbody>
                    {% for key in address %}
                    <tr>
                        <td>{{key.id}}</td>
                        <td>{{key.address}}</td>
                        <td>{{key.phone}}</td>
                        <td><a href="/update_address/{{key.id}}/1/">修改</a></td>
                        <td><a href="/delete_address/{{key.id}}/1/">删除</a></td>
                    </tr>
                    {%endfor%}
                </tbody>
            </table>
            </form><br>
            <form method="get" action="/add_address/1/">
                <input type="submit" value="添加地址">
            </form>
        </div>
    </div>

{%endblock%}
```

模板通过{% extends "base.html" %}和{% block content %}…{% endblock %}套用一个名为"base.html"的模板文件。base.html文件如下。

```html
{%load staticfiles%}
<!DOCTYPE html>
<html lang="zh-CN">
    <head>
        <meta charset="utf-8">
        <meta http-equiv="X-UA-Compatible" content="IE=edge">
        <meta name="viewport" content="width=device-width, initial-scale=1">
        <!--上述3个meta标签*必须*放在最前面,任何其他内容都*必须*跟随其后!-->
        <meta name="description" content="">
        <meta name="author" content="">

        <title>电子商务系统</title>

        <!--Bootstrap core CSS -->
        <link href="{%static 'css/signin.css'%}" rel="stylesheet">
        <!--Custom styles for this template -->
        <link href="{%static 'css/bootstrap.min.css'%}" rel="stylesheet">
    </head>

    <body role="document">
<!--导航栏 -->
    <nav class="navbar navbar-inverse navbar-fixed-top">
        <div class="container">
            <div class="navbar-header">
                <a class="navbar-brand" href="/goods_view/">电子商务系统</a>
            </div>
            <div id="navbar" class="collapse navbar-collapse">
                <ul class="nav navbar-nav">
                    <li class="active"><a href="/goods_view/">商品列表</a></li>
                    <li class="active"><a href="/view_all_order/">查看所有订单</a>
</li>
    {%block content %}


    {%endblock %}
        <div class="container theme-showcase" role="main">

            <div class="page-header">
            </div>
            <footer class="footer">
                <p>&copy; Company 2017,作者:顾翔</p>
            </footer>
```

```
        </div><!--/container -->

<!--Bootstrap core JavaScript
================================================== -->
<!--Placed at the end of the document so the pages load faster -->
<script src="//cdn.bootcss.com/jquery/1.11.3/jquery.min.js"></script>
<script src="//cdn.bootcss.com/bootstrap/3.3.5/js/bootstrap.min.js"></script>

</body>

</html>
```

可以看到，这有点像传统 HTML 中的 frame 或者 iframe，但是使用起来方便得多。

再回到 view_user.html 的页面，views.py 返回当前用户信息的列表参数 user_info，通过{{user_info.username}}和{{user_info.email}}显示当前用户的用户名及 Email 信息。通过如下 HTML 代码修改密码。

```
...
<form action="/change_password/" method="get">
<input type="submit" value="修改密码">
</form>
...
```

进入修改密码的界面。views.py 返回页面当前用户收货地址信息的列表参数 address，由于一个用户可以有一到多个收货地址信息，所以模板通过{% for key in address %}…{% endfor %}在＜table＞…＜/table＞中显示，并且可以对收货地址记录进行修改和删除操作。最后，页面通过下面代码添加当前用户的收货信息。

```
...
<form method="get" action="/add_address/1/">
<input type="submit" value="添加地址">
</form>
...
```

用户信息界面如图 3-4 所示。

用户名：linda Email:xianggu625@126.com 修改密码				
编号	地址	电话	修改	删除
17	上海市闵行区宝城路158弄17号101室	13681732596	修改	删除
18	上海国际会议中心	13681732596	修改	删除
添加地址				

图 3-4　用户信息界面

这里,"<a href="/update_address/{{key.id}}/1/">修改</a>"和"<a href="/delete_address/{{key.id}}/1/">删除</a>"中{{key.id}}/后面的"1"表示收货地址的修改和删除操作在用户信息中进行,从3.6节中的介绍会发现删除和修改操作也可以从生成订单的时候在选择地址中进行操作,那时参数由1改为2。同样,<form method="get" action="/add_address/1/">中的1也是这个道理。

4. 接口测试

1)测试用例

表3-2为用户信息显示模块的测试用例。测试程序测试产品代码是否能够将当前登录用户的信息正确地显示出来。

表 3-2　用户信息显示模块的测试用例

编号	描　　　述	期望结果
1	显示当前登录用户的信息	用户信息正确地显示

2)测试代码及优化

这里将对测试代码进一步封装及优化,这样可以使以后的测试代码维护变得更加简单、灵活。在这个系统的所有测试用例中,均要先在数据库中建立一个新用户,然后用这个用户进行操作,最后在测试结束的时候删除这个用户。所以,利用这个用户的信息专门建立一个XML文件,命名为initInfo.xml,内容如下:

```
<?xml version="1.0" encoding="UTF-8"?>
<node>
    <case>
        <!--初始化用户信息-->
        <id>0</id><!--用户id-->
        <username>Johnson</username><!--用户名称-->
        <password>000000</password><!--用户密码-->
        <email>Johnson@126.com</email><!--Email地址-->
    </case>
</node>
```

其中,

(1)<id>与</id>间的数字为新建用户的id,由于Django后台主键自增变量默认从1开始,所以在这里赋值为0,以避免冲突。

(2)<username>与</username>间的字符串是建立用户的用户名。

(3)<password>与</password>间的字符串是建立用户的密码。

(4)<email>与</email>间的字符串是建立用户的Email地址。

建立测试项目中的interface/util.py。在interface/util.py中,GetXML类及getxmldata()方法改造为如下形式。

```
#!/usr/bin/env python
#coding:utf-8
```

```
import sqlite3,requests
from xml.dom import minidom

class GetXML:
        #获取 xml 中的数据
        def getxmldata(self,myXmlFile):
                dom =minidom.parse(myXmlFile)
                self.root =dom.documentElement
                #从 XML 中读取数据
                TestIds =self.root.getElementsByTagName('TestId')
...
```

然后重新定义 getUserInitInfo()方法,具体代码如下。

```
        #获取 initInfo 中的初始化用户的初始化数据
        def getUserInitInfo(self):
                dom =minidom.parse("initInfo.xml")
                self.root =dom.documentElement
                #从 XML 中读取数据
                id =self.root.getElementsByTagName('id')
                id = (str(id[0].firstChild.data)).strip()
                username =self.root.getElementsByTagName('username')
                username ="\""+ (str(username[0].firstChild.data)).strip()+"\""
                password =self.root.getElementsByTagName('password')
                password ="\""+ (str(password[0].firstChild.data)).strip()+"\""
                email =self.root.getElementsByTagName('email')
                email ="\""+ (str(email[0].firstChild.data)).strip()+"\""
                values =id +","+username+","+password+","+email
                return values     #返回的字符串 values 在插入数据库表 goos_user 中使用
```

userInfoConfig.xml 中的内容如下。

```xml
<?xml version="1.0" encoding="UTF-8"?>
<node>
    <case>
        <login>1</login>
    </case>
    <!--显示用户信息 -->
    <case>
        <TestId>userInfo-testcase001</TestId>
        <Title>用户的显示</Title>
        <Method>get</Method>
        <Desc>显示用户信息</Desc>
        <Url>http://127.0.0.1:8000/user_info/</Url>
        <InptArg></InptArg>
```

```
            <Result>200</Result>
            <CheckWord>Johnson</CheckWord><!--初始化文件中的用户名 -->
        </case>
</node>
```

其中，<login>1</login>中的 1 表示需要登录后再操作，0 表示不需要登录后再操作，所以，如果表示不需要登录后再操作（如用户注册和登录），就可以使用<login>0</login>。在用户信息显示中，程序必须先登录，才能查看登录用户的信息，根据 views.py 中的控制，页面将自动跳转到 login 页面，并且系统会给出"请登录后再进入"的错误提示信息。

```
…
util = Util()
    username = util.check_user(request)
    if username=="":
        uf = LoginForm()
        return render(request,"index.html",{'uf':uf,"error":"请登录后再进入"})
    else:
…
```

然后在 interface/util.py 中建立一个名为 getIsLogin()的方法，作用是获取测试 XML 文件中是否需要登录的信息，内容如下。

```
…
#获取测试 XML 文件中是否需要登录的信息
        def getIsLogin(self,myXmlFile):
            dom = minidom.parse(myXmlFile)
            self.root = dom.documentElement
            #从 XML 中读取数据
            login = self.root.getElementsByTagName('login')
            login = (str(login[0].firstChild.data)).strip()
            return login
…
```

现在建立测试代码 loginTest.py，首先介绍 setUp()方法。

```
…
class userTest(unittest.TestCase):
        def setUp(self):
            print("--------测试开始--------")
            xmlfile ="userInfoConfig.xml"
            #建立 GetXML 对象变量
            xmlInfo = GetXML()
            #获得是否需要登录的信息
```

```
        self.sign = xmlInfo.getIsLogin(xmlfile)
        #获得所有测试数据
        self.mylists = xmlInfo.getxmldata(xmlfile)
        #定义数据库表名
        self.userTable = "goods_user"
        #获得初始化信息
        self.userValues = xmlInfo.getUserInitInfo()
        #建立 DB 对象变量
        self.dataBase = DB()
        #连接数据库
        self.dataBase.connect()
        #插入测试需要的用户
        self.dataBase.insert(self.userTable, self.userValues)
...
```

(1) 通过语句 xmlfile = "userInfoConfig.xml"定义数据驱动读取的测试初始化信息所在的 XML 文件名。

(2) 通过语句 xmlInfo = GetXML()从 XML 文件中获取测试初始化信息。

(3) 通过语句 self.sign = xmlInfo.getIsLogin(xmlfile)获取测试是否需要登录操作，把它赋值给变量 self.sign(0 为不用登录，1 为需要登录)。

(4) 通过语句 self.mylists = xmlInfo.getxmldata(xmlfile)获取所有测试数据，把它赋值给变量 self.mylists，在测试方法中使用。

(5) 通过语句 self.userTable = "goods_user"定义用户数据库表名。

(6) 通过语句 self.userValues = xmlInfo.getUserInitInfo()获得初始化数据库信息。

(7) 建立数据库，把用户信息插入数据库中。

这里把数据库的操作进行封装，在 interface/util.py 中的 Util 类中定义一个名为 insertTable()的方法。

```
#插入数据
#dataBase 为数据库
#table 为数据表
    #values 为值
    def insertTable(dataBase, table, values):
        #连接数据库
        dataBase.connect()
        #插入测试需要的数据
        dataBase.insert(table, values)
```

接下来优化通过 requests 执行接口测试的方法。在 interface/util.py 中的 Util 类中定义一个 run_test()方法，代码如下。

```
        #运行测试接口
        #mylist 为测试数据
        #values 为登录数据
```

```python
def run_test(mylist,values,sign):
    #初始化 requests.Session 变量
    s=requests.session()
    #获取登录数据
    username=values.split(',')[1].strip("\"")
    password=values.split(',')[2].strip("\"")
    #获取测试 URL
    Login_url="http://127.0.0.1:8000/login_action/"
    #login_Url 为登录的 URL
    run_url=mylist["Url"]
    #run_url 为运行测试用例所需的 URL
    #判断当前测试是否需要登录
    if sign:
        #使用当前用户登录系统
        payload={"username":username,"password":password}
        try:
            data=s.post(Login_url,data=payload)
        except Exception as e:
            print(e)
    #运行测试接口
    try:
    #为 POST 请求,由于 POST 请求参数是肯定存在的,所以这里不判断有无参数
        if mylist["Method"]=="post":
            payload=eval(mylist["InptArg"])
            data=s.post(run_url,data=payload)
        #为 GET 请求,需要判断有无参数
        elif mylist["Method"]=="get":
            if mylist["InptArg"].strip()=="":
                #没有请求参数
                data=s.get(run_url)
            else:
                #有请求参数
                payload=eval(mylist["InptArg"])
                data=s.get(run_url,params=payload)
    except Exception as e:
        print(e)
    else:
        return data
```

（1）通过语句 s = requests.session()初始化 requests.session 变量。

（2）获取接口测试如果需要登录,就在初始化设置中设置登录用户名和密码,对应代码为 username = values.split(',')[1].strip("\"")和 password = values.split(',')[2].strip("\"")。

（3）设置运行路径 run_url = mylist["Url"]和登录路径 Login_url = "http://127.0.

0.1:8000/login_action/"。

(4) 通过标记 sign 判断是否需要登录,如果需要登录,就调用 payload ={"username":username,"password":password}以及 data = s.post(Login_url,data=payload)语句,然后正式进入接口测试环节。

(5) 通过判断语句 if mylist["Method"] == "post":判断请求为 POST 方法,调用 payload = eval(mylist["InptArg"])与 data = s.post(run_url,data=payload)语句。

(6) 否则,请求方式为 GET,分别处理有参数与没有参数两种情形。

① 当有参数的时候,调用 payload = eval(mylist["InptArg"])和 data = s.get(run_url,params=payload)语句。

② 当没有参数的时候,调用 data = s.get(run_url)语句。

这样,测试程序的主体部分就变得非常简单了,代码如下。

```python
#开始测试
    def test_user_info(self):
        for mylist in self.mylists:
            data=self.util.run_test(mylist,self.userValues,self.sign)
            #验证返回码
            self.assertEqual(mylist["Result"],str(data.status_code))
            #验证返回文本
            self.assertIn(mylist["CheckWord"],str(data.text))
            print (mylist["TestId"]+" is passing!")
```

(1) 通过循环语句 for mylist in self.mylists:遍历所有测试用例。

(2) 调用上面提及的 run_test 方法 data = self.util.run_test(mylist,self.userValues, self.sign)。

(3) 通过语句 self.assertEqual(mylist["Result"],str(data.status_code))判断返回码是否正确。

(4) 通过语句 self.assertIn(mylist["CheckWord"],str(data.text))验证文本是否在返回文本中。

(5) 最后,通过语句 print (mylist["TestId"]+" is passing!")打上这个测试用例通过的标记。

tearDown()方法定义如下。

```python
def tearDown(self):
    #获取初始化数据库中的记录主码
    id=self.userValues.split(',')[0]
    #删除这条记录
    self.dataBase.delete(self.userTable,"id="+id)
    #关闭数据库连接
    self.dataBase.close()
    print("--------测试结束--------")
```

封装这个方法,在 interface/util.py 中的 util 类中定义一个方法 tearDown()。

```python
def tearDown(dataBase,table,values,sign):
    #获取初始化数据库中的记录主码
    id=values.split(',')[0]
    #删除这条记录
    dataBase.delete(table,"id="+id)
    #关闭数据库连接
    dataBase.close()
```

这样，在 userTest.py 中使用起来就非常简单了。

```python
def tearDown(self):
            Util.tearDown(self.dataBase,self.userTable,self.userValues,self.sign)
            print("--------测试结束--------")
```

这里把用户注册与用户登录也进行改造，并且加上注释信息。

登录配置文件 loginConfig.xml。

```xml
<?xml version="1.0" encoding="UTF-8"?>
<node>
    <!--登录操作,正确用户名,错误密码 -->
    <case>
        <TestId>loginReg-testcase001</TestId>
        <Title>用户登录</Title>
        <Method>post</Method>
        <Desc>正确用户名,错误密码</Desc>
        <Url>http://127.0.0.1:8000/login_action/</Url>
        <InptArg>{"username":"Johnson","password":"123456"}</InptArg><!--密码与 initInfo.xml 中用户信息的密码不同 -->
        <Result>200</Result>
        <CheckWord>用户名或者密码错误</CheckWord>
    </case>
    <!--登录操作,错误用户名,正确密码 -->
    <case>
        <TestId>loginReg-testcase002</TestId>
        <Title>用户登录</Title>
        <Method>post</Method>
        <Desc>错误用户名,正确密码</Desc>
        <Url>http://127.0.0.1:8000/login_action/</Url>
        <InptArg>{"username":"smith","password":"000000"}</InptArg><!--用户名与 initInfo.xml 中用户信息的用户名不同 -->
        <Result>200</Result>
        <CheckWord>用户名或者密码错误</CheckWord>
    </case>
    <!--登录操作,错误用户名,错误密码 -->
```

```xml
<case>
    <TestId>loginReg-testcase003</TestId>
    <Title>用户登录</Title>
    <Method>post</Method>
    <Desc>错误用户名,错误密码</Desc>
    <Url>http://127.0.0.1:8000/login_action/</Url>
    <InptArg>{"username":"smith","password":"123456"}</InptArg><!--用户
名、密码与 initInfo.xml 中用户信息的用户名、密码不同 -->
    <Result>200</Result>
    <CheckWord>用户名和密码错误</CheckWord>
</case>
<!--登录操作成功 -->
<case>
    <TestId>loginReg-testcase004</TestId>
    <Title>用户登录</Title>
    <Method>post</Method>
    <Desc>正确用户名,正确密码</Desc>
    <Url>http://127.0.0.1:8000/login_action/</Url>
    <InptArg>{"username":"Johnson","password":"000000"}</InptArg><!--用户
名、密码与 initInfo.xml 中用户信息的用户名、密码相同 -->
    <Result>200</Result>
    <CheckWord>查看购物车</CheckWord>
</case>
</node>
```

登录测试代码 loginTest.py。

```python
#!/usr/bin/env python
#coding:utf-8
import unittest,requests
from util import GetXML,DB,Util

class mylogin(unittest.TestCase):
    def setUp(self):
        print("--------测试开始--------")
        xmlfile = "loginConfig.xml"
        #建立 GetXML 对象变量
        xmlInfo = GetXML()
        #获得是否需要登录的信息
        self.sign = xmlInfo.getIsLogin(xmlfile)
        #获得所有测试数据
        self.mylists = xmlInfo.getxmldata(xmlfile)
        #定义数据库表名
        self.userTable = "goods_user"
```

```python
            #获得初始化信息
            self.userValues = xmlInfo.getUserInitInfo()
            #建立DB对象变量
            self.dataBase = DB()
            Util.insertTable(self.dataBase,self.userTable, self.userValues)

    def test_login(self):
        for mylist in self.mylists:
            data = Util.run_test(mylist, self.userValues, self.sign)
            #验证返回码
            self.assertEqual(mylist["Result"],str(data.status_code))
            #验证返回文本
            self.assertIn(mylist["CheckWord"],str(data.text))

    def tearDown(self):
        Util.tearDown(self.dataBase, self.userTable, self.userValues, self.sign)
        print("--------测试结束--------")

if __name__ == '__main__':
    #构造测试集
    suite=unittest.TestSuite()
    suite.addTest(mylogin("test_login"))
    #运行测试集合
    runner=unittest.TextTestRunner()
    runner.run(suite)
```

注册配置文件 registerConfig.xml。

```xml
<?xml version="1.0" encoding="UTF-8"?>
<node>
    <case>
        <login>0</login><!--0表示执行不需要登录、1表示需要登录 -->
    </case>
    <!--注册的时候,用户名已经存在 -->
    <case>
        <TestId>loginReg-testcase001</TestId>
        <Title>用户注册</Title>
        <Method>post</Method>
        <Desc>注册的时候,用户名已经存在</Desc>
        <Url>http://127.0.0.1:8000/register/</Url>
        <InptArg>{"username":"Johnson","password":"000000","email":"Johnson5@126.com"}</InptArg><!--用户名同initInfo.xml中用户信息的用户名 -->
```

```xml
        <Result>200</Result>
        <CheckWord>用户名已经存在!</CheckWord>
    </case>
    <!--注册成功 -->
    <case>
        <TestId>loginReg-testcase002</TestId>
        <Title>用户注册</Title>
        <Method>post</Method>
        <Desc>注册用户名不存在</Desc><!--通过"注册用户名不存在"测试完毕,删除数据库记录 -->
        <Url>http://127.0.0.1:8000/register/</Url>
        <InptArg>{"username":"smith","password":"000000","email":"smith@126.com"}</InptArg><!--用户名与 initInfo.xml 中用户信息的用户名不同 -->
        <Result>200</Result>
        <CheckWord>登录</CheckWord>
    </case>
</node>
```

注册测试代码 registerTest.py。

```python
#!/usr/bin/env python
#coding:utf-8
import unittest,requests
from util import GetXML,DB,Util

class myregister(unittest.TestCase):

    def setUp(self):
        print("--------测试开始--------")
        xmlfile = "registerConfig.xml"
        #建立 GetXML 对象变量
        xmlInfo = GetXML()
        #获得是否需要登录的信息
        self.sign = xmlInfo.getIsLogin(xmlfile)
        #获得所有测试数据
        self.mylists = xmlInfo.getxmldata(xmlfile)
        #定义数据库表名
        self.userTable = "goods_user"
        #获得初始化信息
        self.userValues = xmlInfo.getUserInitInfo()
        #建立 DB 对象变量
        self.dataBase = DB()
        Util.insertTable(self.dataBase,self.userTable,self.userValues)
```

```python
        def test_register(self):
            for mylist in self.mylists:
                data = Util.run_test(mylist, self.userValues, self.sign)
                #验证返回码
                self.assertEqual(mylist["Result"], str(data.status_code))
                #验证返回文本
                self.assertIn(mylist["CheckWord"], str(data.text))
                #如果注册成功,就删除刚建立的记录
                if "注册用户名不存在" in mylist["Desc"]:
                    payload = eval(mylist["InptArg"])
                    username = "\""+payload["username"]+"\""
                    self.dataBase.delete(self.userTable, "username="+username)

        def tearDown(self):
            Util.tearDown(self.dataBase, self.userTable, self.userValues, self.sign)
            print("--------测试结束--------")

if __name__ == '__main__':
    #构造测试集
    suite = unittest.TestSuite()
    suite.addTest(myregister("test_register"))
    #运行测试集合
    runner = unittest.TextTestRunner()
    runner.run(suite)
```

接下来,再对测试程序及配置文件进行优化,把 loginConfig.xml 和 registerConfig.xml 这两个文件进行合并,形成 loginRegConfig.xml。

```xml
<?xml version="1.0" encoding="UTF-8"?>
<node>
    <case>
        <login>0</login><!-- 0 表示执行不需要登录、1 表示需要登录 -->
    </case>
    <!-- 注册的时候,用户名已经存在 -->
    <case>
        <TestId>loginReg-testcase001</TestId>
        <Title>用户注册</Title>
        <Method>post</Method>
        <Desc>注册的时候,用户名已经存在</Desc>
        <Url>http://127.0.0.1:8000/register/</Url>
        <InptArg>{"username":"Johnson","password":"000000","email":"Johnson5@126.com"}</InptArg><!-- 用户名同 initInfo.xml 中用户信息的用户名 -->
        <Result>200</Result>
```

```xml
            <CheckWord>用户名已经存在!</CheckWord>
        </case>
        <!--注册成功 -->
        <case>
            <TestId>loginReg-testcase002</TestId>
            <Title>用户注册</Title>
            <Method>post</Method>
            <Desc>注册用户名不存在</Desc><!--通过"注册用户名不存在"测试完毕,删除数据库记录 -->
            <Url>http://127.0.0.1:8000/register/</Url>
            <InptArg>{"username":"smith","password":"000000","email":"smith@126.com"}</InptArg><!--用户名与initInfo.xml中用户信息的用户名不同-->
            <Result>200</Result>
            <CheckWord>登录</CheckWord>
        </case>
        <!--登录操作,正确用户名,错误密码 -->
        <case>
            <TestId>loginReg-testcase003</TestId>
            <Title>用户登录</Title>
            <Method>post</Method>
            <Desc>正确用户名,错误密码</Desc>
            <Url>http://127.0.0.1:8000/login_action/</Url>
            <InptArg>{"username":"Johnson","password":"123456"}</InptArg><!--密码与initInfo.xml中用户信息的密码不同-->
            <Result>200</Result>
            <CheckWord>用户名或者密码错误</CheckWord>
        </case>
        <!--登录操作,错误用户名,正确密码 -->
        <case>
            <TestId>loginReg-testcase004</TestId>
            <Title>用户登录</Title>
            <Method>post</Method>
            <Desc>错误用户名,正确密码</Desc>
            <Url>http://127.0.0.1:8000/login_action/</Url>
            <InptArg>{"username":"smith","password":"000000"}</InptArg><!--用户名与initInfo.xml中用户信息的用户名不同-->
            <Result>200</Result>
            <CheckWord>用户名或者密码错误</CheckWord>
        </case>
        <!--登录操作,错误用户名,错误密码 -->
        <case>
            <TestId>loginReg-testcase005</TestId>
            <Title>用户登录</Title>
            <Method>post</Method>
```

```xml
        <Desc>错误用户名,错误密码</Desc>
        <Url>http://127.0.0.1:8000/login_action/</Url>
        <InptArg>{"username":"smith","password":"123456"}</InptArg><!--用户
名、密码与 initInfo.xml 中用户信息的用户名、密码不同-->
        <Result>200</Result>
        <CheckWord>用户名和密码错误</CheckWord>
    </case>
    <!--登录操作成功 -->
    <case>
        <TestId>loginReg-testcase006</TestId>
        <Title>用户登录</Title>
        <Method>post</Method>
        <Desc>正确用户名,正确密码</Desc>
        <Url>http://127.0.0.1:8000/login_action/</Url>
        <InptArg>{"username":"Johnson","password":"000000"}</InptArg><!--用户
名、密码与 initInfo.xml 中用户信息的用户名、密码相同-->
        <Result>200</Result>
        <CheckWord>查看购物车</CheckWord>
    </case>
</node>
```

然后合并 loginTest.py 和 registerTest.py,形成 loginRegTest.py。

```python
#!/usr/bin/env python
#coding:utf-8
import unittest,requests
from util import GetXML,DB,Util

class loginreg(unittest.TestCase):

    def setUp(self):
        print("--------测试开始--------")
        xmlfile = "loginRegConfig.xml"
        #建立 GetXML 对象变量
        xmlInfo = GetXML()
        #获得是否需要登录的信息
        self.sign = xmlInfo.getIsLogin(xmlfile)
        #获得所有测试数据
        self.mylists = xmlInfo.getxmldata(xmlfile)
        #定义数据库表名
        self.userTable = "goods_user"
        #获得初始化信息
        self.values = xmlInfo.getUserInitInfo()
        #建立 DB 对象变量
```

```
            self.dataBase =DB()
            Util.insertTable(self.dataBase,self.userTable,self.values)

    def test_register(self):
        for mylist in self.mylists:
            data =Util.run_test(mylist,self.values,self.sign)
            #验证返回码
            self.assertEqual(mylist["Result"],str(data.status_code))
            #验证返回文本
            self.assertIn(mylist["CheckWord"],str(data.text))
            #如果注册成功,就删除刚建立的记录
            if "注册用户名不存在" in mylist["Desc"]:
                payload =eval(mylist["InptArg"])
                username ="\""+payload["username"]+"\""
                self.dataBase.delete(self.userTable,"username="+username)
            print (mylist["TestId"]+" is passing!")

    def tearDown(self):
        Util.tearDown(self.dataBase,self.userTable,self.values)
        print("--------测试结束--------")

if __name__=='__main__':
    #构造测试集
    suite=unittest.TestSuite()
    suite.addTest(myregister("test_register"))
    #运行测试集合
    runner=unittest.TextTestRunner()
    runner.run(suite)
```

在这里要加上下面几行代码。

```
if "注册用户名不存在" in mylist["Desc"]:
    payload =eval(mylist["InptArg"])
    username ="\""+payload["username"]+"\""
    self.dataBase.delete(self.userTable,"username="+username)
```

上述几条语句的作用是:在测试注册最后,对于注册成功的记录进行删除操作,其目的是保证每个测试程序间的互相独立性。也就是说,其他测试用例不受这条测试用例的影响,这在自动化测试中非常重要。

刚才在 interface/util.py 中封装了 tearDown(),现在封装 setUp()。在 Util 类中建立一个 inivalue() 方法。

```
#初始化信息
    def inivalue(self,dataBase,ordertable,sign):
```

```
            #获得初始化信息
            if sign =="0":
                    values =GetXML.getUserInitInfo(self)
            elif sign =="1":
                    values =GetXML.getGoodInitInfo(self)
            elif sign =="2":
                    values =GetXML.getAddressInitInfo(self)
            elif sign =="3":
                    values =GetXML.getOrdersInitInfo(self)
            elif sign =="4":
                    values =GetXML.getOrderInitInfo(self)
            else:
                    print("sign is error in function inivalue")
            #建立记录
            self.insertTable(dataBase,ordertable,values)
            return values
```

考虑到下面模块的使用，把用户、商品、收货地址、单个订单和总订单的初始化均写在这里了。

这样，loginRegTest.py 改造完成。

```
#!/usr/bin/env python
#coding:utf-8
import unittest,requests
from util import GetXML,DB,Util

class myregister(unittest.TestCase):

        def setUp(self):
                print("--------测试开始--------")
                xmlfile ="loginRegConfig.xml"
                #建立 GetXML 对象变量
                xmlInfo =GetXML()
                #获得是否需要登录的信息
                self.sign =xmlInfo.getIsLogin(xmlfile)
                #获得所有测试数据
                self.mylists =xmlInfo.getxmldata(xmlfile)
                #建立 DB 变量
                self.dataBase =DB()
                #建立 util 变量
                self.util =Util()
                #初始化用户记录
                self.userTable ="goods_user"
```

```
            self.userValues = self.util.inivalue(self.dataBase,self.userTable,"0")
            ...
```

userTest.py 改动完成。

```python
#!/usr/bin/env python
#coding:utf-8
import unittest,requests
from util import GetXML,DB,Util

class userTest(unittest.TestCase):
    def setUp(self):
        print("--------测试开始--------")
        xmlfile ="userInfoConfig.xml"
        #建立 GetXML 对象变量
        xmlInfo =GetXML()
        #获得是否需要登录的信息
        self.sign =xmlInfo.getIsLogin(xmlfile)
        #获得所有测试数据
        self.mylists =xmlInfo.getxmldata(xmlfile)
        #建立 DB 变量
        self.dataBase =DB()
        #建立 util 变量
        self.util =Util()
        #初始化用户记录
        self.userTable ="goods_user"
        self.userValues =self.util.inivalue(self.dataBase,self.userTable,"0")
...
```

可以看出，以后的接口测试代码可以变得非常简单。这样，接口测试工作主要是设计测试用例和书写 XML 文件（即设计测试数据），而不是维护测试代码，因为大部分代码已经被封装了。最后建立一个 runtest.py 文件，内容如下。

```python
#!/usr/bin/env python
#coding:utf-8
import unittest
from HTMLTestRunner import HTMLTestRunner

test_dir='./'
discover=unittest.defaultTestLoader.discover(test_dir,pattern="*Test.py")

if __name__=='__main__':
    runner=unittest.TextTestRunner()
    #以下用于生成测试报告
    fp=open("result.html","wb")
```

```
runner=HTMLTestRunner(stream=fp,title='测试报告',description='测试用例执行
报告')
    runner.run(discover)
    fp.close()
```

运行这个测试代码,形成测试报告,如图 3-5 所示。

图 3-5　测试报告

以后测试代码均以 Test.py 结尾,就可以使用这个 TestSuite 文件了。

3.3.4　用户登录密码的修改

系统为用户提供用户登录密码的修改。根据需求定义,修改用户密码的时候,必须提供旧密码、新密码和新密码的确认信息,并且新密码不能与旧密码相同。如果旧密码不正确、新密码与旧密码相同或者新密码和新密码的确认信息不一致,系统应该给出相应的提示信息。

1. urls.py

```
...
url(r'^change_password/$',views.change_password),
...
```

2. views.py

```
...
#修改用户密码
def change_password(request):
    util=Util()
    username=util.check_user(request)
    if username=="":
        uf=LoginForm()
        return render(request,"index.html",{'uf':uf,"error":"请登录后再进入"})
    else:
        count=util.cookies_count(request)
```

```
            #获得当前登录用户的用户信息
            user_info =get_object_or_404(User,username=username)
            #如果是提交表单,就获取表单信息,并且进行表单信息验证
            if request.method =="POST":
                #获取旧密码
                oldpassword= (request.POST.get("oldpassword", "")).strip()
                #获取新密码
                newpassword= (request.POST.get("newpassword", "")).strip()
                #获取新密码的确认密码
                checkpassword= (request.POST.get("checkpassword", "")).strip()
                #如果旧密码不正确,就报错误信息,不允许修改
                if oldpassword !=user_info.password:
                    return render(request,"change_password.html",{"user": username,
"error":"旧密码不正确","count":count})
                #如果新密码与旧密码相同,就报错误信息,不允许修改
                elif newpassword ==oldpassword:
                    return render(request,"change_password.html",{"user": username,
"error":"新密码不能与旧密码相同","count":count})
                #如果新密码与新密码的确认密码不匹配,就报错误信息,不允许修改
                elif newpassword !=checkpassword:
                    return render(request,"change_password.html",{"user": username,
"error":"确认密码与新密码不匹配","count":count})
                else:
                    #否则修改成功
                    User.objects.filter(username=username).update(password=newpassword)
                    return render(request,"change_password.html",{"user": username,
"error":"密码修改成功","count":count})
            #如果不是提交表单,就显示修改密码页面
            else:
                return render (request,"change_password.html",{"user": username,
"count":count})
        ...
```

(1) 首先确认当前用户是否为登录用户。

(2) 然后判断是否为表单提交状态。如果不是,就显示修改密码页面,否则获取用户输入的旧密码、新密码和新密码确认密码。

(3) 最后进行如下三项判断操作。

① 旧密码是否正确。

② 新密码与旧密码是否不相同。

③ 新密码与新密码的确认密码是否相同。

(4) 如果不满足需求,就跳转到 change_password.html 页面中显示错误信息,否则通过代码 User.objects.filter(username=username).update(password=newpassword)保存新密码。

(5) 返回 change_password.html 显示密码修改正确的信息。

3. 模板

change_password.html：

```html
{%extends "base.html" %}
{%block content %}
            <li class="active"><a href="/view_chart/">查看购物车<font color="#FF0000">{{ count }}</font></a></li>
          </ul>
          <ul class="nav navbar-nav navbar-right">
            <li><a href="/user_info/">{{user}}</a></li>
            <li><a href="/logout/">退出</a></li>
          </ul>
        </div><!--/.nav-collapse -->
      </div>
    </nav>


    <div class="container theme-showcase" role="main">

      <div class="page-header">
        <div id="navbar" class="navbar-collapse collapse">
        </div><!--/.navbar-collapse -->
      </div>

      <div class="row">
        <div class="col-md-6">
          <form action ="/change_password/" method =" post" enctype="multipart/form-data">
              <p style="color:red">{{error}}</p><br>
              旧密码:<input name="oldpassword" type="password" required><br>
              新密码:<input name="newpassword" type="password" required><br>
              确认密码:<input name="checkpassword" type="password" required><br>
              <input type="submit" value="修改">
          </form>
        </div>

      </div>

{%endblock %}
```

用户密码修改页面如图3-6所示。

4. 接口测试

1) 测试用例

表3-3为修改用户密码测试用例。假设旧密码为"000000"，新密码为"123456"，设计4个测试用例，分别如下。

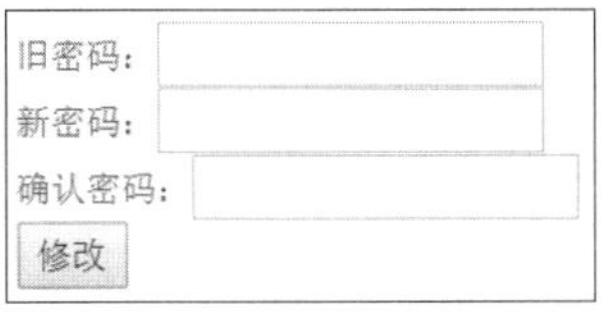

图 3-6 用户密码修改页面

（1）旧密码错误，提示错误信息"旧密码不正确"。
（2）新密码与旧密码相同，提示错误信息"新密码不能与旧密码相同"。
（3）确认密码与新密码不匹配，提示错误信息"确认密码与新密码不匹配"。
（4）旧密码、确认密码与新密码设置正确，显示"密码修改成功"的信息。

表 3-3 修改用户密码测试用例

编号	描述			期望结果
	旧密码	新密码	新密码的确认密码	
1	123456	654321	654321	提示"旧密码不正确"
2	000000	000000	000000	提示"新密码不能与旧密码相同"
3	000000	123456	654321	提示"确认密码与新密码不匹配"
4	000000	123456	123456	显示"密码修改成功"的信息

2）XML 数据文件

根据测试用例的设计，在测试配置文件 userInfoConfig.xml 中加入如下内容。

```xml
...
    <!--修改用户密码,旧密码不正确 -->
    <case>
        <TestId>userInfo-testcase002</TestId>
        <Title>修改用户密码</Title>
        <Method>post</Method>
        <Desc>旧密码不正确</Desc>
        <Url>http://127.0.0.1:8000/change_password/</Url>
        <InptArg>{"oldpassword":"123456","newpassword":"654321","checkpassword":"654321"}</InptArg><!--旧密码与初始化密码不相同 -->
        <Result>200</Result>
        <CheckWord>旧密码不正确</CheckWord>
    </case>
    <!--修改用户密码,新密码与旧密码相同 -->
    <case>
        <TestId>userInfo-testcase003</TestId>
        <Title>修改用户密码</Title>
        <Method>post</Method>
        <Desc>新密码不能与旧密码相同</Desc>
        <Url>http://127.0.0.1:8000/change_password/</Url>
```

```xml
            <InptArg>{"oldpassword":"000000","newpassword":"000000", "checkpassword":
"000000"}</InptArg><!--新密码与旧密码相同 -->
            <Result>200</Result>
            <CheckWord>新密码不能与旧密码相同</CheckWord>
    </case>
    <!--修改用户密码,确认密码与新密码不匹配 -->
    <case>
        <TestId>userInfo-testcase004</TestId>
        <Title>修改用户密码</Title>
        <Method>post</Method>
        <Desc>确认密码与新密码不匹配</Desc>
        <Url>http://127.0.0.1:8000/change_password/</Url>
        <InptArg>{"oldpassword":"000000","newpassword":"123456","checkpassword":
"654321"}</InptArg><!--确认密码与新密码不匹配 -->
        <Result>200</Result>
        <CheckWord>确认密码与新密码不匹配</CheckWord>
    </case>
    <!--修改用户密码,密码修改成功 -->
    <case>
        <TestId>userInfo-testcase005</TestId>
        <Title>修改用户密码</Title>
        <Method>post</Method>
        <Desc>密码修改成功</Desc>
        <Url>http://127.0.0.1:8000/change_password/</Url>
        <InptArg>{"oldpassword":"000000","newpassword":"123456","checkpassword":
"123456"}</InptArg><!--新密码与旧密码不同,确认密码与新密码不匹配 -->
        <Result>200</Result>
        <CheckWord>密码修改成功</CheckWord>
    </case>
</node>
```

3）测试代码

这里,测试代码不需要做任何变化。读者有没有发现,前面对测试代码进行了比较好的封装和优化,使得接口测试工作变得更加简单。

3.4 商品信息模块

商品信息模块包括"商品信息的维护""商品概要信息的分页显示""根据商品名称的模糊查询"和"对某一条商品显示其详细信息"。商品信息的维护通过Django提供的后台进行操作。

数据模型如下。

```
...
#商品
class Goods(models.Model):
    name =models.CharField(max_length=100)              #商品名称
    price =models.FloatField()                          #单价
    picture =models.FileField(upload_to ='./upload/')   #图片
    desc =models.TextField()                            #描述

    def __str__(self):
        return self.name
...
```

3.4.1 商品信息的维护

商品信息的维护包括商品信息的添加、修改和删除。由于 Django 提供了相当庞大的后台管理模块,所以,对商品信息进行维护就使用 Django 提供的后台。

通过 http://127.0.0.1/admin/进入 Django 提供的后台,找到 Goodss 一行,如图 3-7 所示。

图 3-7 商品信息维护界面

单击图标 ➕ Add 进入图 3-8,添加商品信息。

图 3-8 添加商品信息

单击图标 ✏ Change 进入图 3-9,显示商品信息列表页面。

勾选复选框选中几条商品信息,从下拉列表中选中 Delete selected goodss ▾ ,然后单击按钮 Go ,删除选择的商品信息,如图 3-10 所示。

图 3-9　显示商品信息列表页面

图 3-10　删除选择的商品信息

单击商品名称的链接，就可以修改这条商品信息的记录，如图 3-11 所示。

图 3-11　修改商品信息的记录

3.4.2 商品概要信息的分页显示

商品概要信息的分页显示页面是登录操作以后的首界面,以列表的形式显示已经存在的商品,通过这个页面,用户可以查看商品的概要信息,添加商品进入购物车等。

1. urls.py

```
...
url(r'^goods_view/$', views.goods_view),
...
```

2. views.py

```
...
#查看商品信息
def goods_view(request):
    util =Util()
    username =util.check_user(request)
    if username =="":
        uf =LoginForm()
        return render(request,"index.html",{'uf':uf,"error":"请登录后再进入"})
    else:
        #获得所有商品信息
        good_list =Goods.objects.all()
        #获得购物车中的物品数量
        count =util.cookies_count(request)

        #翻页操作
        paginator =Paginator(good_list, 5)
        page =request.GET.get('page')
        try:
            contacts =paginator.page(page)
        except PageNotAnInteger:
            #如果页号不是一个整数,就返回第一页
            contacts =paginator.page(1)
        except EmptyPage:
            #如果页号超出范围(如 9999),就返回结果的最后一页
            contacts =paginator.page(paginator.num_pages)
        return render(request, "goods_view.html", {"user": username, "goodss": contacts,"count":count})
...
```

(1) 通过代码 good_list = Goods.objects.all()获得所有的商品信息。

(2) 通过 paginator = Paginator(good_list,5)以及下面的代码实现分页显示的功能。

3. 模板

goods_view.html：

```html
{% extends "base.html" %}
{% block content %}
            <li class="active"><a href="/view_chart/">查看购物车<font color="#FF0000">{{ count }}</font></a></li>
          </ul>
          <ul class="nav navbar-nav navbar-right">
            <li><a href="/user_info/">{{user}}</a></li>
            <li><a href="/logout/">退出</a></li>
          </ul>
        </div><!--/.nav-collapse -->
      </div>
    </nav>


    <div class="container theme-showcase" role="main">

        <!--商品表单-->
        <div class="page-header">
            <div id="navbar" class="navbar-collapse collapse">
                <form class="navbar-form" method="post" action="/search_name/">
                    <div class="form-group">
                        <input name="good" type="text" placeholder="名称" class="form-control">
                    </div>
                    <button type="submit" class="btn btn-success">搜索</button>
                </form>
            </div><!--/.navbar-collapse -->
        </div>

        <div class="row">
            <div class="col-md-6">
                <table class="table table-striped">
                    <thead>
                        <tr>
                            <th>编号</th>
                            <th>名称</th>
                            <th>价格</th>
                            <th>查看详情</th>
                            <th>放入购物车</th>
                        </tr>
                    </thead>
                    <tbody>
                        {% for goods in goodss %}
```

```
                    <tr>
                        <td>{{ goods.id }}</td>
                        <td>{{ goods.name }}</td>
                        <td> ¥ {{ goods.price }}</td>
                        <td><a href="/view_goods/{{goods.id}}/">查看</a>
</td>
                        <td><a href="/add_chart/{{goods.id}}/1/">放入</a>
</td>
                    </tr>
                    {%endfor %}
                </tbody>
            </table>
        </div>

    </div>
<!--列表分页器 -->
    <div class="pagination">
        <span class="step-links">
            {%if goodss.has_previous %}
                <a href="?page={{ goodss.previous_page_number }}">上一页</a>
            {%endif %}
            <span class="current">
                Page {{ goodss.number }} of {{ goodss.paginator.num_pages }}.
            </span>
            {%if goodss.has_next %}
                <a href="?page={{ goodss.next_page_number }}">下一页</a>
            {%endif %}
        </span>
    </div>
{%endblock %}
```

所有商品通过列表变量 goodss 返回给模板文件,在模板文件中通过{% for goods in goodss %}遍历显示。分页功能通过模板中<!--列表分页器 -->下面的代码实现,如图 3-12 所示。

4. 接口测试

1) 测试用例

表 3-4 为商品信息列表的测试用例。测试目的是把测试数据中的商品信息插入数据库中,检验这个商品的列表信息是否可以正确地被显示出来。

表 3-4　商品信息列表的测试用例

编号	描　　述	期望结果
1	分页显示当前所有商品的概要信息	添加的信息能够被及时地显示出来

编号	名称	价格	查看详情	放入购物车
1	正山堂茶业 元正简雅正山小种红茶茶叶礼盒装礼品 武夷山茶叶送礼	¥238.0	查看	放入
2	红茶茶叶 正山小种武夷山红茶170g 春茶袋装170g 散装新茶	¥25.0	查看	放入
3	晋袍 花蜜香正山小种红茶 300g牛皮纸袋装礼盒 武夷山桐木关包邮	¥188.0	查看	放入
4	春茶 茶叶 桐木关红茶 正山小种 野生小种茶叶直销 品牌	¥368.0	查看	放入
5	正山小种红茶特级 新茶 礼盒装 桂圆香 送礼红茶暖养胃茶叶250g	¥238.12	查看	放入

Page 1 of 2. 下一页

图 3-12　商品信息列表

2) XML 文件

首先在 initInfo.xml 中加入初始化数据。

```
...
<!--初始化商品信息 -->
    <case>
        <goodid>0</goodid><!--商品 id -->
        <name>龙井茶叶</name><!--商品名称 -->
        <price>1234.56</price><!--商品价格 -->
        <picture>upload/0.jpg</picture><!--商品图片 -->
        <desc>龙井茶叶龙井茶叶龙井茶叶龙井茶叶龙井茶叶龙井茶叶龙井茶叶龙井茶叶龙井茶叶龙井茶叶龙井茶叶龙井茶叶龙井茶叶龙井茶叶</desc><!--商品描述 -->
    </case>
...
```

然后建立测试数据配置文件 goodsConfig.xml,内容如下。

```
<?xml version="1.0" encoding="UTF-8"?>
<node>
    <case>
        <login>1</login>
    </case>
    <!--翻页显示当前所有的商品信息 -->
    <case>
        <TestId>goods-testcase001</TestId>
        <Title>商品信息</Title>
        <Method>get</Method>
```

```xml
        <Desc>显示商品列表信息</Desc>
        <Url>http://127.0.0.1:8000/goods_view/</Url>
        <InptArg></InptArg>
        <Result>200</Result>
        <CheckWord>龙井茶叶</CheckWord><!-- 与初始化商品名称保持一致 -->
    </case>
</node>
```

3) 测试代码

首先,在interface/util.py文件GetXML中建立一个方法getGoodInitInfo(),用于从初始化initInfo.xml文件中获取建立goods表的内容,便于在测试程序setUp()方法中使用。

```python
#获取initInfo.xml中的商品初始化数据
    def getGoodInitInfo(self):
        dom = minidom.parse("initInfo.xml")
        self.root = dom.documentElement
        #从XML中读取数据
        goodid = self.root.getElementsByTagName('goodid')
        goodid = (str(goodid[0].firstChild.data)).strip()
        name = self.root.getElementsByTagName('name')
        name = "\""+(str(name[0].firstChild.data)).strip()+"\""
        price = self.root.getElementsByTagName('price')
        price = (str(price[0].firstChild.data)).strip()
        picture = self.root.getElementsByTagName('picture')
        picture = "\""+(str(picture[0].firstChild.data)).strip()+"\""
        desc = self.root.getElementsByTagName('desc')
        desc = "\""+(str(desc[0].firstChild.data)).strip()+"\""
        values = goodid+","+name+","+price+","+picture+","+desc
        return values    #返回的字符串values供插入数据库表goods_goods中使用
```

然后在这个文件中的Util类中加入方法tearDown(),用于删除setUp中建立的good信息。

```python
def tearDown(self,dataBase,table,values):
    #获取初始化数据库中的记录主码
    id = values.split(',')[0]
    #删除这条记录
    dataBase.delete(table,"id="+id)
```

建立测试代码goodsInfoTest.py。

```python
#!/usr/bin/env python
#coding:utf-8
import unittest,requests
from util import GetXML,DB,Util
```

```python
class goodTest(unittest.TestCase):
    def setUp(self):
        print("--------测试开始--------")
        xmlfile = "goodsConfig.xml"
        #建立 GetXML 对象变量
        xmlInfo = GetXML()
        #获得是否需要登录的信息
        self.sign = xmlInfo.getIsLogin(xmlfile)
        #获得所有测试数据
        self.mylists = xmlInfo.getxmldata(xmlfile)
        #建立 DB 变量
        self.dataBase = DB()
        #建立 util 变量
        self.util = Util()
        #初始化用户记录
        self.userTable = "goods_user"
        self.userValues = self.util.inivalue(self.dataBase,self.userTable,"0")
        #初始化商品记录
        self.goodsTable = "goods_goods"
        self.goodsValues = self.util.inivalue(self.dataBase,self.goodsTable,"1")

    #开始测试
    def test_goods_info(self):
        for mylist in self.mylists:
            data = self.util.run_test(mylist,self.userValues,self.sign)
            #验证返回码
            self.assertEqual(mylist["Result"],str(data.status_code))
            #验证返回文本
            self.assertIn(mylist["CheckWord"],str(data.text))
            print (mylist["TestId"]+" is passing!")

    def tearDown(self):
        self.util.tearDown(self.dataBase,self.goodsTable,self.goodsValues)
        self.util.tearDown(self.dataBase,self.userTable,self.userValues)
        #关闭数据库连接
        self.dataBase.close()
        print("--------测试结束--------")

if __name__ == '__main__':
    #构造测试集
    suite=unittest.TestSuite()
    suite.addTest(goodTest("test_goods_info"))
    #运行测试集合
    runner=unittest.TextTestRunner()
    runner.run(suite)
```

在方法 setUp() 中,建立用户后,通过下面的语句建立初始化商品信息。

```
#初始化商品记录
self.goodsTable ="goods_goods"
self.goodsValues =self.util.inivalue(self.dataBase,self.goodsTable,"1")
```

最后在 tearDown() 方法中通过下列语句删除在 setUp() 方法中建立的商品信息。

```
self.util.tearDown(self.dataBase,self.goodsTable,self.goodsValues)
```

同样,使用下面的语句删除在 setUp() 方法中建立的用户信息。

```
self.util.tearDown(self.dataBase,self.userTable,self.userValues)
```

这里对 interface/util.py 中 Util 类里的 insertTable() 方法进行优化。假设在上一个测试中出现了错误,测试程序没有正常结束,即没有完成 tearDown() 方法中的代码就进行下一个测试用例的执行,这样就可能出现在这次 setUp() 方法中往数据库表中插入的数据由于上次没有执行 tearDown() 方法中的删除记录操作而重复插入,导致这次插入不成功。所以,对方法 insertTable() 进行如下改动。

```
...
#插入数据
    #dataBase 为数据库
    #table 为数据库表
    #values 为值
    def insertTable(dataBase,table,values):
        #获取插入数据的 id
        id =values.split(',')[0].strip("\"")
        #连接数据库
        dataBase.connect()
        #查询数据库表中是否存在这条记录
        if dataBase.searchByid(table,id):
            #如果存在,就删除这条记录
            dataBase.delete(table,"id="+id)
        #插入测试需要的用户
        dataBase.insert(table,values)
...
```

粗体字部分为新加的内容,同时在这里新加了一个 searchByid() 方法,其代码如下。

```
#通过主键查询数据库表中的内容
def searchByid(self,tablename,id):
    return(self.cur.execute("select * from "+tablename+" where id="+id))
```

在 insertTable() 方法中,在插入数据库表前,通过语句 if dataBase.searchByid(table,id) 先判断数据库表中是否存在这条记录,如果存在,就删除它。

3.4.3 商品信息的模糊查询

1. urls.py

```
...
url(r'^search_name/$', views.search_name),
...
```

2. views.py

```python
...
#商品搜索
def search_name(request):
    util = Util()
    username = util.check_user(request)
    if username == "":
        uf = LoginForm()
        return render(request,"index.html",{'uf':uf,"error":"请登录后再进入"})
    else:
        count = util.cookies_count(request)
        #获取查询数据
        search_name = (request.POST.get("good", "")).strip()
        #通过 objects.filter()方法进行模糊匹配查询,查询结果放入变量 good_list
        good_list = Goods.objects.filter(name__icontains=search_name)

        #对查询结果进行分页显示
        paginator = Paginator(good_list, 5)
        page = (request.GET.get('page')).strip()
        try:
            contacts = paginator.page(page)
        except PageNotAnInteger:
            #如果页号不是一个整数,就返回第一页
            contacts = paginator.page(1)
        except EmptyPage:
            #如果页号超出范围(如 9999),就返回结果的最后一页
            contacts = paginator.page(paginator.num_pages)
        return render(request, "goods_view.html", {"user": username, "goodss": contacts,"count":count})
...
```

这里的实现方法与商品概要信息基本一致,不同的地方是,在概要信息中使用代码 good_list = Goods.objects.all()获取全部商品信息,而在模糊查询中使用代码 good_list = Goods.objects.filter(name__icontains=search_name)显示符合条件的商品信息。

3. 模板

这里的模板同商品信息列表的模板。

4．接口测试

1）测试用例

表 3-5 为商品信息模糊搜索的测试用例，这里设计了 3 个测试用例。

（1）正常的测试用例，查询数据库中符合条件的商品信息，系统应该把这个商品信息正确地显示出来。

（2）查询字符为空的字符串，系统应该把所有的数据都显示出来。

（3）主要检验模糊查询中是否存在 SQL 注入，在查询字符中输入 SQL 模糊查询通配符"％"，系统应该显示商品标题中含有"％"的商品，由于测试程序中不含有"％"的商品，所以查询结果应该为空。

表 3-5 商品信息模糊搜索的测试用例

编号	输入数据	期望结果
1	目前已知商品名称的一部分	这个商品信息被查询且显示出来
2	空字符	显示所有内容
3	％	不显示所有内容

2）XML 文件

在 goodsConfig.xml 中加入如下内容。

```xml
...
    <!--输入数据:目前已知商品名称的子串,期望结果:这个商品信息被查询且显示出来 -->
    <case>
        <TestId>goods-testcase003</TestId>
        <Title>商品信息</Title>
        <Method>post</Method>
        <Desc>查询商品</Desc>
        <Url>http://127.0.0.1:8000/search_name/</Url>
        <InptArg>{"good":"龙井"}</InptArg>
        <Result>200</Result>
        <CheckWord>龙井</CheckWord><!--包含查询商品名称的一部分 -->
    </case>
    <!--输入数据:空字符,期望结果:显示所有内容 -->
    <case>
        <TestId>goods-testcase004</TestId>
        <Title>商品信息</Title>
        <Method>post</Method>
        <Desc>查询商品</Desc>
        <Url>http://127.0.0.1:8000/search_name/</Url>
        <InptArg>{"good":""}</InptArg>
        <Result>200</Result>
        <CheckWord>龙井茶叶</CheckWord><!--与初始化商品名称保持一致 -->
    </case>
```

```xml
<!--输入数据:%,期望结果:不显示所有内容 -->
<case>
    <TestId>goods-testcase005</TestId>
    <Title>商品信息</Title>
    <Method>post</Method>
    <Desc>查询商品</Desc>
    <Url>http://127.0.0.1:8000/search_name/</Url>
    <InptArg>{"good":"%"}</InptArg>
    <Result>200</Result>
    <CheckWord>NOT,龙井茶叶</CheckWord><!--表示初始化商品信息的名称在响应信息中不被显示,这里的"NOT"表示不显示 -->
</case>
...
```

最后一个测试数据中,<CheckWord>NOT,龙井茶叶</CheckWord>表示"龙井茶叶"没有被查询出来,其中 NOT 表示不显示。

3)测试代码

根据<CheckWord>NOT,龙井茶叶</CheckWord>表示"龙井茶叶"没有被查询出来,修改测试代码 goodsInfoTest.py 中的 test_goods_info()方法。

```python
...
#开始测试
        def test_goods_info(self):
                for mylist in self.mylists:
                        data=self.util.run_test(mylist,self.userValues,self.sign)
                        #验证返回码
                        self.assertEqual(mylist["Result"],str(data.status_code))
                        #验证返回文本
#如果 mylist["CheckWord"]标签中存在"NOT"字符串,就调用断言方法 assertNotIn()
if "NOT" in mylist["CheckWord"]:
                        self.assertNotIn((mylist["CheckWord"]).split(",")[1],str(data.text))
                        #否则调用断言方法 assertIn()
                        else:
                                self.assertIn(mylist["CheckWord"],str(data.text))
print (mylist["TestId"]+" is passsing!")
...
```

请注意粗体字部分;这里不再进一步介绍了。

3.4.4 商品信息的详情显示

查看商品信息详情的实现方法与查看商品信息列表的实现方法基本相同。

1. urls.py

```
...
url(r'^view_goods/(?P<good_id>[0-9]+)/$', views.view_goods),
...
```

这里,r'^view_goods/(?P<good_id>[0-9]+)/$'表示 view_goods/后面跟着一个由数字组成的字符串,这个字符串定义为变量 good_id,供 views.py 中使用。good_id 为商品信息的 id。

2. views.py

```
...
#查看商品信息详情
def view_goods(request,good_id):
    util =Util()
    username =util.check_user(request)
    if username=="":
        uf =LoginForm()
        return render(request,"index.html",{'uf':uf,"error":"请登录后再进入"})
    else:
        count =util.cookies_count(request)
        good =get_object_or_404(Goods, id=good_id)
        return render(request, 'good_details.html', {"user": username,'good':good,"count":count})
...
```

程序通过语句 good = get_object_or_404(Goods,id=good_id)获取所要显示商品的详细信息,然后通过变量 good 传递给模板显示。

3. 模板

good_details.html

```
{%extends "base.html" %}
{%block content %}
                <li class="active"><a href="/view_chart/">查看购物车<font color="#FF0000">{{ count }}</font></a></li>
            </ul>
            <ul class="nav navbar-nav navbar-right">
                <li><a href="/user_info/">{{user}}</a></li>
                <li><a href="/logout/">退出</a></li>
            </ul>
        </div><!--/.nav-collapse -->
    </div>
</nav>

<div class="container theme-showcase" role="main">
```

```html
<!--商品详细信息-->
<div class="row" style="margin-top:30px">
    <div class="col-lg-6">
        <div class="input-group">
            单价：¥ {{good.price}}元<a href="/add_chart/{{good.id}}/2/">放入购物车</a><br>
            <img src="/{{good.picture}}"><br>
            {{good.desc}}
        </div><!--/input-group -->
    </form>
    </div><!--/.col-lg-6 -->
</div><!--/.row -->

</div><!--/container    glyphicon glyphicon-phone border-style:none; -->

{%endblock %}
```

通过代码<img src="/{{good.picture}}">直接显示图片，再次强调""""后紧跟着一个"/"，不要忘记。商品详细信息显示如图3-13所示。

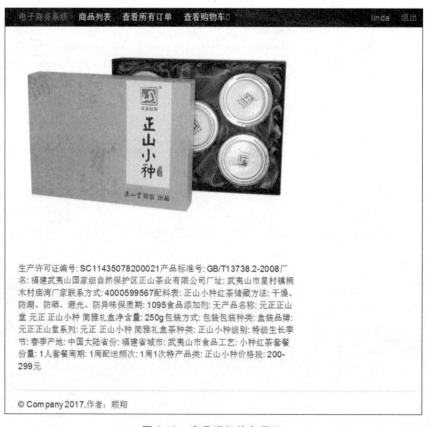

图3-13 商品详细信息显示

4. 接口测试

1）测试用例

表 3-6 为商品详细信息测试用例。测试目的是把测试数据中的商品信息插入数据库，检验这个商品的详细信息是否可以正确地被显示出来。

表 3-6 商品详细信息测试用例

编号	描　　述	期望结果
1	显示当前商品的详细信息	当前的商品信息被正确地显示出来

2）XML 文件

这里仍旧使用 initInfo.xml 加入初始化商品数据。在测试数据文件 goodsConfig.xml 中增加如下内容。

```
...
<!--显示当前商品的详细信息 -->
    <case>
        <TestId>goods-testcase002</TestId>
        <Title>商品信息</Title>
        <Method>get</Method>
        <Desc>显示商品的详细信息</Desc>
        <Url>http://127.0.0.1:8000/view_goods/0/</Url>
        <InptArg></InptArg>
        <Result>200</Result>
        <CheckWord>龙井茶叶龙井茶叶龙井茶叶龙井茶叶龙井茶叶龙井茶叶龙井茶叶龙井茶叶龙井茶叶龙井茶叶龙井茶叶龙井茶叶龙井茶叶</CheckWord><!--与初始化商品的详细信息保持一致 -->
    </case>
...
```

3）测试代码

这里，接口测试的代码不需要做任何改动。

3.5　购物车模块

购物车模块包括"把商品放入购物车""查看购物车中的商品""修改购物车中的商品数量""删除购物车中的某种商品"和"删除购物车内所有的商品"。

在程序中采取 cookie 的形式存储购物车中的内容，大家都知道，一个 cookie 是一个值参对，在参数中存放商品的 id，通过商品的 id 从数据库中查询对应的商品信息。在值中存放商品的数量，初始化时为 1，然后在查看购物车中的内容页面中提供修改购物车内商品数量的功能。

购物车模块不具有数据模型。

3.5.1 把商品放入购物车

1. urls.py

```
...
url(r'^add_chart/(?P<good_id>[0-9]+)/(?P<sign>[0-9]+)/$', views.add_chart),
...
```

(1) good_id 为商品的 id。
(2) sign=1 表示从商品列表添加进的商品,sign=2 表示从商品详情添加进的商品。

2. views.py

```
...
#放入购物车
def add_chart(request,good_id,sign):
    util =Util()
    username =util.check_user(request)
    if username=="":
        uf =LoginForm()
        return render(request,"index.html",{"error":"请登录后再进入"})
    else:
        #获得商品详情
        good =get_object_or_404(Goods,id=good_id)
        #如果 sign=="1",则返回商品列表页面
        if sign=="1":
            response =HttpResponseRedirect('/goods_view/')
        #否则返回商品详情页面
        else:
            response =HttpResponseRedirect('/view_goods/'+good_id)
        #把当前商品添加进购物车,参数为商品 id,值为购买商品的数量,默认为 1,有效时间为一年
        response.set_cookie(str(good.id),1,60*60*24*365)
        return response
...
```

(1) 注册用户登录后,通过语句 good = get_object_or_404(Goods,id=good_id)获取添加进购物车中商品的信息变量。

(2) 通过语句 response.set_cookie(str(good.id),1,60*60*24*365)把商品放入购物车,实现方式是利用 cookie,其中 cookie 的 key 为商品的 id,即 str(good.id)、购物车中商品的初始数量默认为 1,有效时间为一年,即 60*60*24*365(单位为秒)。

在 goods/util.py 中,cookies_count()方法的具体代码如下。

```
...
#返回购物车内商品的个数
def cookies_count(self,request):
```

```
#返回本地所有的cookies
cookie_list = request.COOKIES
#只要进入网站,系统中就会产生一个名为sessionid的cookie
#如果后台同时在运行,就会产生一个名为csrftoken的cookie
if "csrftoken" in cookie_list:
    return len(request.COOKIES)-2
else:
    return len(request.COOKIES)-1
...
```

1.3.5节中已经介绍了sessionId,现在再回顾一下。sessionId是一个会话的key,目的是解决原始HTTP的无状态性。浏览器在第一次访问服务器的时候会在服务器端生成一个session,并且有一个sessionId与它对应。不同的Web服务器对于sessionId有不同的解释,例如,Tomcat生成的sessionId叫作jsessionId。当客户端第一次请求session对象时,服务器会为客户端创建一个session,并将通过特殊算法算出一个session的ID,用来标识该session对象。另外,还有一个系统可能产生的cookie名为csrftoken,当setting.py中启动CSRF防御的时候,这个cookie就会产生。所以,代码在计算购物车中商品数量的时候要将这两个cookie去除,其中sessionId的cookie是肯定会产生的,所以直接去除就可以了,而csrftoken的cookie在需要开启CSRF防御的时候才会产生,所以通过条件语句if "csrftoken" in cookie_list进行判断。

3. 模板

放入购物车后,在页面上会自动更新显示购物车中商品的数量,这里不需要提供专门的模板,在任何登录后的菜单栏中均可以显示。如图3-14所示,菜单"查看购物车"右边的2就是商品的数量。

图3-14 显示购物车内商品的数量

4. 接口测试

1) 测试用例

表3-7为放入购物车的测试用例。通过商品列表页面或者商品详情页面把商品放入购物车,购物车内的商品数量会相应增加。

表3-7 放入购物车的测试用例

编号	描述	期望结果
1	添加一个商品到购物车	添加成功,购物车内的商品数量会相应增加

这里,测试思路如下。

(1) 初始化建立的用户登录信息。

(2) 为了保证测试的有效性,删除购物车中的所有记录。

(3) 把初始化建立的商品信息添加到购物车中。

(4) 检查显示购物车商品数量是否为 1。

2) XML 数据文件

chartConfig.xml:

```xml
<?xml version="1.0" encoding="UTF-8"?>
<node>
    <case>
        <login>1</login>
    </case>
    <!--添加商品到购物车,查看显示购物车内商品数量的变化 -->
    <case>
        <TestId>chart-testcase001</TestId>
        <Title>购物车</Title>
        <Method>get</Method>
        <Desc>添加进购物车</Desc>
        <Url>http://127.0.0.1:8000/add_chart/0/1/</Url>
        <InptArg></InptArg>
        <Result>200</Result>
        <CheckWord>查看购物车 &lt;font color=&quot;#FF0000&quot;&gt;1&lt;/font&gt;&lt;/a&gt;</CheckWord>
    </case>
```

注意,XML 中的特殊字符需要进行转义,进行转义的字符同 HTML,总结如下。

(1) &(逻辑与)——>&。

(2) <(小于)——><。

(3) >(大于)——>>。

(4) "(双引号)——>"。

(5) '(单引号)——>'。

转义的时候需要注意以下几方面。

(1) 转义序列各字符间不能有空格。

(2) 转义序列必须以";"结束。

(3) 单独的"&"不被认为是转义开始。

(4) 注意区分大小写。

这里,"查看购物车 <font color="#FF0000">1</font></a>"为"查看购物车<font color="#FF0000">1</font></a>"。

3) 测试代码

chartinfoTest.py:

```python
#!/usr/bin/env python
#coding:utf-8
```

```python
import unittest,requests
from util import GetXML,DB,Util

class charttest(unittest.TestCase):
    def setUp(self):
        print("--------测试开始--------")
        xmlfile = "chartConfig.xml"
        #建立 GetXML 对象变量
        xmlInfo = GetXML()
        #获得是否需要登录的信息
        self.sign = xmlInfo.getIsLogin(xmlfile)
        #获得所有测试数据
        self.mylists = xmlInfo.getxmldata(xmlfile)
        #建立 DB 变量
        self.dataBase = DB()
        #建立 util 变量
        self.util = Util()
        #初始化用户记录
        self.userTable = "goods_user"
        self.userValues = self.util.inivalue(self.dataBase,self.userTable,"0")
        #初始化商品记录
        self.goodTable = "goods_goods"
        self.goodValues = self.util.inivalue(self.dataBase,self.goodTable,"1")

    #开始测试
    def test_chart_info(self):
        #初始化购物车,把购物车中的所有内容均删除
        data = self.util.initChart()
        for mylist in self.mylists:
            data = self.util.run_test(mylist,self.userValues,self.sign)
            #验证返回码
            self.assertEqual(mylist["Result"],str(data.status_code))
            self.assertIn(mylist["CheckWord"],str(data.text))
            print (mylist["TestId"]+" is passing!")

    def tearDown(self):
        #删除在测试过程中在购物车中加入的商品信息
        self.util.tearDownByCookie()
        #删除 setup 建立的商品
        self.util.tearDown(self.dataBase,self.goodTable,self.goodValues)
        #删除 setup 建立的用户
        self.util.tearDown(self.dataBase,self.userTable,self.userValues)
        #关闭数据库连接
        self.dataBase.close()
```

```
                print("--------测试结束--------")
if __name__=='__main__':
    #构造测试集
    suite=unittest.TestSuite()
    suite.addTest(charttest("test_chart_info"))
    #运行测试集合
    runner=unittest.TextTestRunner()
    runner.run(suite)
```

注意两处粗体字部分,进行测试时,通过 Util.initChart()方法清除购物车中的所有商品记录。在 interface/util.py 中,initChart()方法的实现代码如下。

```
def initChart():
    s = requests.session()
    data = s.get("http://127.0.0.1:8000/remove_chart_all/")
```

这里调用产品代码 view.py 中的 remove_chart_all()方法实现。注意,当测试完毕后,一定要把在测试中建立的购物车中的商品信息删除。同样,在 interface/util.py 中,tearDownByCookie()调用产品代码 views.py 中的 remove_chart()方法实现。

```
def tearDownByCookie():
    s = requests.session()
    data = s.get("http://127.0.0.1:8000/remove_chart/0/")
```

在这里再对测试程序进行进一步的优化。现在测试与开发是在一台机器上进行的,所以这里用到的地址均为 http://127.0.0.1:8000,为了测试代码的易维护性,可把它作为 Util 类的成员变量,在初始化的时候声明。

```
...
class Util:
    def __init__(self):
        self.url = "http://127.0.0.1:8000"
...
```

然后在下面几个地方就可以使用了。

```
...
Login_url = self.url+"/login_action/"    #Login_Url 为登录的 URL
...
data = s.get(self.url+"/remove_chart_all/")
...
data = s.get(self.url+"/remove_chart/0/")
...
```

并且,interface/util.py 的 Util 类中的所有方法都加上参数 self。

```
    ...
    def insertTable(self,dataBase,table,values):
    ...
    def run_test(self,mylist,values,sign):
    ...
    def initChart(self):
    ...
    def tearDownByCookie(self):
    ...
    def tearDown(self,dataBase,table,values):
    ...
```

经过上述通用代码的调整,在具体的测试代码中要进行如下修改,这里仅以 goodsInfoTest.py 为例。

```
            ...
            self.util =Util()
            self.util.insertTable(self.dataBase,self.userTable,self.userValues)
            #定义商品数据库表名
            self.Goodstable = "goods_goods"
            #获得商品初始化信息
            self.goodvalues = xmlInfo.getGoodInitInfo()
            #建立商品记录
            self.util.insertTable(self.dataBase,self.Goodstable,self.goodvalues)

    #开始测试
    def test_goods_info(self):
            for mylist in self.mylists:
                    data = self.util.run_test(mylist,self.userValues,self.sign)
                    ...

    def tearDown(self):
            self.util.tearDown(self.dataBase,self.Goodstable,self.goodvalues)
            self.util.tearDown(self.dataBase,self.userTable,self.userValues)
            print("--------测试结束--------")
    ...
```

代码中的粗体字部分为具体的调整方法。

3.5.2 查看购物车中的商品

1. urls.py

```
...
url(r'^view_chart/$', views.view_chart),
...
```

2. views.py

```
...
#查看购物车
def view_chart(request):
    util =Util()
    username=util.check_user(request)
    if username=="":
        u =LoginForm()
        return render(request,"index.html",{'uf':uf,"error":"请登录后再进入"})
    else:
        #购物车中的商品个数
        count=util.cookies_count(request)
        #返回所有的cookie内容
        my_chart_list =util.add_chart(request)
        return render(request, "view_chart.html", {"user": username, "goodss": my_chart_list, "count":count})
...
```

登录用户通过调用语句 my_chart_list = util. add_chart(request)把商品放入购物车中。在产品代码 Util 类中的 add_chart()方法代码如下。

```
...
#加入购物车
    def add_chart(self,request):
        #获取购物车内的所有内容
        cookie_list =self.deal_cookies(request)
        #定义 my_chart_list 列表
        my_chart_list =[]
        #遍历 cookie_list,把里面的内容加入类 Chart_list 的 my_chart_list 列中
        for key in cookie_list:
            chart_object =Chart_list
            chart_object =self.set_chart_list(key,cookie_list)
            my_chart_list.append(chart_object)
        #返回 my_chart_list
        return my_chart_list
...
```

(1) 调用方法 deal_cookies()获取购物车内的所有内容。

(2) 通过语句 for key in cookie_list 遍历 cookie_list。

(3) 通过方法 set_chart_list()把 cookie_list 里的内容加入类 Chart_list 的 my_chart_list 列表变量中。

deal_cookies()方法如下。

```
...
    #获取购物车内的所有内容
    def deal_cookies(self,request):
        #获取本地所有的COOKIES
        cookie_list = request.COOKIES
        #去除COOKIES内的sessionid
        cookie_list.pop("sessionid")
        #如果COOKIES内含有csrftoken,则去除COOKIES内的csrftoken
        if "csrftoken" in cookie_list:
            cookie_list.pop("csrftoken")
        #否则返回处理好的购物车内的所有内容
        return cookie_list
...
```

方法 set_chart_list()用于把获取购物车中的商品放在一个名为 Chart_list 的类中,返回模板,其代码如下。

```
...
#把购物车中的商品放在一个名为Chart_list()的类中,返回模板
def set_chart_list(self,key,cookie_list):
    chart_list = Chart_list()
    good_list = get_object_or_404(Goods, id=key)
    chart_list.set_id(key)                        #商品的id
    chart_list.set_name(good_list.name)           #商品的名称
    chart_list.set_price(good_list.price)         #商品的价格
    chart_list.set_count(cookie_list[key])        #商品的数量
    return chart_list
...
```

(1) 通过语句 good_list = get_object_or_404(Goods,id=key)获得商品信息。

(2) 通过语句 chart_list.set_id(key)、chart_list.set_name(good_list.name)、chart_list.set_price(good_list.price)和 chart_list.set_count(cookie_list[key])分别把商品的 id、名称、价格及数量放入 Chart_list 类中。

关于 Chart_list 类,系统在 object.py 中如下定义。

```
...
#购物车模型
class Chart_list():
    #主键
    def set_id(self,id):
        self.id=id

    #商品的名称
    def set_name(self,name):
```

```
        self.name=name

    #商品的价格
    def set_price(self,price):
        self.price=price

    #商品的数量
    def set_count(self,count):
        self.count=count
...
```

在 object.py 中定义类模型,除了上面提及的购物车模型类 Chart_list,还包括订单中的订单模型类 Order_list 及总订单模型类 Orders_list。

3. 模板

view_chart.html:

```
...
{% extends "base.html" %}
{% block content %}
                </ul>
                <ul class="nav navbar-nav navbar-right">
                    <li><a href="/user_info/">{{user}}</a></li>
                    <li><a href="/logout/">退出</a></li>
                </ul>
            </div><!--/.nav-collapse -->
        </div>
    </nav>
    <div class="page-header">
        <div id="navbar" class="navbar-collapse collapse">
        </div><!--/.navbar-collapse -->
    </div>
    <div class="container theme-showcase" role="main">
    <font color="#FF0000">{{error}}</font>
        <div class="row">
            <div class="col-md-6">
                <table class="table table-striped">
                    <thead>
                        <tr>
                            <th>编号</th>
                            <th>名称</th>
                            <th>价格</th>
                            <th>数量</th>
                            <th>移除</th>
                        </tr>
```

```
                </thead>
                <tbody>
                    {% for key in goodss %}
                        <tr>
                            <td>< a href="/view_goods/{{key.id}}/">{{key.id}}</a></td>
                            <td>{{key.name}}</td>
                            <td>¥ {{key.price}}</td>
                            <td>< form action="/update_chart/{{key.id}}/" method="post"><input type="number" value="{{key.count}}" style="width:30px;" name="count{{key.id}}" class="vIntegerField" id="id_count" required /><input type="submit" value="修改" /></form></td>
                            <td><a href="/remove_chart/{{key.id}}/">移除</a></td>
                        </tr>
                    {%endfor%}
                </tbody>
            </table>
            <a href="/remove_chart_all/">清除所有</a>    <a href="/view_address/">生成订单</a>
        </div>

    </div>
{%endblock%}...
```

通过{% for key in goodss %}遍历 Chart_list 类，从而显示购物车中的商品。在这里可以实现修改商品的数量、删除某个商品以及删除购物车内的所有商品，如图 3-15 所示。

图 3-15 显示购物车中商品的内容

4. 接口测试

1) 测试用例

表 3-8 为查看购物车中内容的测试用例。前面把商品放入购物车内，这里验证进入购物车的商品信息是否可以正确地被显示出来。

表 3-8　查看购物车中内容的测试用例

编号	描　　述	期望结果
1	添加一个商品到购物车	在购物车里可以查看这个商品

2）XML 数据文件

在 chartConfig.xml 中添加如下代码。

```xml
...
    <!--将商品添加到购物车,在购物车列表页面查看这个商品 -->
    <case>
        <TestId>chart-testcase002</TestId>
        <Title>购物车</Title>
        <Method>get</Method>
        <Desc>查看购物车中的内容</Desc>
        <Url>http://127.0.0.1:8000/view_chart/</Url>
        <InptArg></InptArg>
        <Result>200</Result>
        <CheckWord>&lt;td&gt;龙井茶叶 &lt;/td&gt;</CheckWord><!--购物车中显示了添加的商品 -->
    </case>
...
```

在＜CheckWord＞…＜/CheckWord＞中，"<td>龙井茶叶 </td>"为"<td>龙井茶叶</td>"。

3）测试代码

下面再做一些优化，把变量 s 作为类的成员变量，这样使用到 s 的地方就改为 self.s。

```python
...
    class Util:
        def __init__(self):
            self.url = "http://127.0.0.1:8000"
            self.s = requests.session()
        ...
        def run_test(self,mylist,values,sign):
            ...
            data = self.s.post(Login_url,data=payload)
            ...
            data = self.s.post(run_url)
            ...
            data = self.s.post(run_url,data=payload)
            ...
            data = self.s.get(run_url,params=payload)
            ...
            data = self.s.get(run_url)
```

```
    ...
        def initChart(self):
            data = self.s.get(self.url+"/remove_chart_all/")
        ...
        data = self.s.get(self.url+"/remove_chart/0/")
    ...
```

然后改测试代码 chartTest.py。

```
    ...
    #开始测试
    def chart_test(self):
            #初始化购物车,把购物车中的所有内容均删除
            data = self.util.initChart()
            for mylist in self.mylists:
                    data = self.util.run_test(mylist,self.userValues,self.sign)
                    #验证返回码
                    self.assertEqual(mylist["Result"],str(data.status_code))
                    #验证返回文本
                    self.assertIn(mylist["CheckWord"],str(data.text))
                    print (mylist["TestId"]+" is passing!")
    ...
```

可以看到,写代码时需要不停地进行优化(这里不管是产品代码,还是测试代码),这样可以使以后更好、更方便地调用这些代码,从而使代码达到较高的复用性和较好的易维护性。所以,代码的优化也是通过不断迭代完成的。一开始就完成一个优秀的代码是不太可能的,就像写书一样,也是经过不断调整、优化而完成的。读者在日常工作中除了编写代码以外,在其他方面也要学会使用这种不断迭代优化的方法。

3.5.3 修改购物车中的商品数量

1. urls.py

```
...
url(r'^update_chart/(?P<good_id>[0-9]+)/$', views.update_chart),
...
```

good_id 为购物车中需要修改商品的商品 id。

2. views.py

```
...
#修改购物车中商品的数量
def update_chart(request,good_id):
    util = Util()
    username = util.check_user(request)
```

```
            if username=="":
                uf = LoginForm()
                return render(request,"index.html",{'uf':uf,"error":"请登录后再进入"})
            else:
                #获取编号为good_id的商品
                good = get_object_or_404(Goods, id=good_id)
                #获取修改的数量
                count = (request.POST.get("count"+good_id, "")).strip()
                #如果数量值≤0,就报出错信息
                if int(count)<=0:
                    #获得购物车列表信息
                    my_chart_list = util.add_chart(request)
                    #返回错误信息
                    return render(request, "view_chart.html", {"user": username, "goodss": my_chart_list,"error":"个数不能小于等于0"})
                else:
                    #否则修改商品数量
                    response = HttpResponseRedirect('/view_chart/')
                    response.set_cookie(str(good.id),count,60 * 60 * 24 * 365)
                    return response
        ...
```

(1) 登录的用户通过语句 good = get_object_or_404(Goods, id=good_id)获取修改商品数量的商品信息。

(2) 通过 count = (request.POST.get("count"+good_id, "")).strip()获得修改商品数量的值。

① 如果获取的值小于或者等于零,系统就调用 view_chart.html 模板,报"个数不能小于等于0"的提示信息(由于模板 view_chart.html 中修改商品数量使用的是＜input type = "number"＞数据类型,所以里面的类型肯定为整数类型)。注意,由于要返回 view_chart.html,所以必须通过语句 my_chart_list = util.add_chart(request)获取购物车中的所有商品。

② 否则通过语句 response.set_cookie(str(good.id),count,60 * 60 * 24 * 365)修改指定商品的数量,返回方法 view_chart()。

3. 模板

当用户修改商品数量后,不管填写的数字是否合法,均返回查看购物车页面,所以,这里的模板与"查看购物车"模块是同一模板。

4. 接口测试

1) 测试用例

表 3-9 为修改购物车中商品数量的测试用例。这里设计了 3 个测试用例。第一个测试用例为正常的测试用例,修改数量为 9。由于购物车中数量是不可以小于或等于 0 的,所以设计了第二个和第三个测试用例,分别把个数修改为 0 和 −1,系统应该有相应的报错信息"个数不能小于或等于 0"。

表 3-9 修改购物车中商品数量的测试用例

编号	描述	期望结果
1	修改购物车中的商品数量为 9	修改成功并且正确地显示
2	修改购物车中的商品数量为 0	报错误信息"个数不能小于或等于 0"
3	修改购物车中的商品数量为 －1	报错误信息"个数不能小于或等于 0"

2）XML 数据文件

```xml
...
<!--修改购物车中的商品数量为 9 -->
    <case>
        <TestId>chart-testcase003</TestId>
        <Title>购物车</Title>
        <Method>post</Method>
        <Desc>修改购物车中的商品数量为 9</Desc>
        <Url>http://127.0.0.1:8000/update_chart/0/</Url>
        <InptArg>{"count0":"9"}</InptArg>
        <Result>200</Result>
        <CheckWord>&lt;inputtype=&quot;number&quot; value=&quot;9&quot;</CheckWord>
<!--购物车中显示了商品数量的变更 -->
    </case>
    <!--修改购物车中的商品数量为 0 -->
    <case>
        <TestId>chart-testcase004</TestId>
        <Title>购物车</Title>
        <Method>post</Method>
        <Desc>修改购物车中的商品数量为 0</Desc>
        <Url>http://127.0.0.1:8000/update_chart/0/</Url>
        <InptArg>{"count0":"0"}</InptArg>
        <Result>200</Result>
        <CheckWord>个数不能小于或等于 0</CheckWord><!--验证购物车中商品数量=0 是
不允许的 -->
    </case>
    <!--修改购物车中的商品数量为-1 -->
    <case>
        <TestId>chart-testcase005</TestId>
        <Title>购物车</Title>
        <Method>post</Method>
        <Desc>修改购物车中的商品数量为-1</Desc>
        <Url>http://127.0.0.1:8000/update_chart/0/</Url>
        <InptArg>{"count0":"-1"}</InptArg><!--验证购物车中的商品数量<0 是不允许
的 -->
        <Result>200</Result>
        <CheckWord>个数不能小于或等于 0</CheckWord>
    </case>
```

```
...
```

在修改数量的form表中,输入框的name为"count"+商品id,由于测试数据的商品id均为0,所以输入框的name为count0。

3) 测试代码

测试代码在这里不做任何修改。

3.5.4 删除购物车中的某种商品

1. urls.py

```
...
url(r'^remove_chart/(?P<good_id>[0-9]+)/$', views.remove_chart),
...
```

good_id为购物车中待删除商品的id。

2. views.py

```
...
#把购物车中的商品移出购物车
def remove_chart(request,good_id):
    util = Util()
    username = util.check_user(request)
    if username == "":
        uf = LoginForm()
        return render(request,"index.html",{'uf':uf,"error":"请登录后再进入"})
    else:
        #获取指定id的商品
        good = get_object_or_404(Goods, id=good_id)
        response = HttpResponseRedirect('/view_chart/')
        #移出购物车
        response.set_cookie(str(good.id),1,0)
        return response
...
```

(1) 登录的用户通过语句good = get_object_or_404(Goods, id=good_id)获得需要移出的商品信息。

(2) 通过语句response.set_cookie(str(good.id),1,0)将其移出购物车。移出购物车,只要把cookie的生效时间改为小于或者等于0即可(本处设置为0)。

3. 模板

当指定商品从购物车被删除以后,会返回查看购物车页面,所以这里的模板与"查看购物车"模块是同一模板。

4. 接口测试

1) 测试用例

表3-10为删除购物车中一个商品的测试用例。从购物车中删除指定的商品,检验这个

商品是否不在购物车的商品详情中显示。

表 3-10 删除购物车中一个商品的测试用例

编号	描 述	期望结果
1	把初始化的商品从购物车中删除	删除成功,不在购物车的商品详情中显示

2）XML 数据文件

```
...
<!--把初始化的商品从购物车中删除 -->
  <case>
      <TestId>chart-testcase006</TestId>
      <Title>购物车</Title>
      <Method>get</Method>
      <Desc>把初始化的商品从购物车中删除</Desc>
      <Url>http://127.0.0.1:8000/remove_chart/0/</Url>
      <InptArg></InptArg>
      <Result>200</Result>
      <CheckWord>NOT,龙井茶叶</CheckWord><!--龙井茶叶在购物车中不显示 -->
  </case>
...
```

与商品中的测试用例 goods-testcase005 一样,CheckWord 中的"NOT,龙井茶叶",表示"龙井茶叶"不在购物车中。

3）测试代码

由于现在的<CheckWord>标签中出现了 NOT,所以在原有的测试代码 chartTest.py 基础上,按照商品测试代码进行修改。

```
...
    #如果 mylist["CheckWord"]标签中存在"NOT"字符串,则调用断言方法 assertNotIn()
    if "NOT" in mylist["CheckWord"]:
        self.assertNotIn((mylist["CheckWord"]).split(",")[1],str(data.text))
    #否则调用断言方法 assertIn()
    else:
        self.assertIn(mylist["CheckWord"],str(data.text))
...
```

3.5.5 删除购物车内所有的商品

1. **urls.py**

```
...
url(r'^remove_chart_all/$', views.remove_chart_all),
...
```

2. views.py

```
...
#删除购物车内所有的商品
def remove_chart_all(request):
    util = Util()
    username = util.check_user(request)
    if username == "":
        uf = LoginForm()
        return render(request,"index.html",{'uf':uf,"error":"请登录后再进入"})
    else:
        response = HttpResponseRedirect('/view_chart/')
        #获取购物车中的所有商品
        cookie_list = util.deal_cookies(request)
        #遍历购物车中的商品,一个一个地删除
        for key in cookie_list:
            response.set_cookie(str(key),1,0)
        return response
...
```

(1) 登录的用户通过语句 cookie_list = util.deal_cookies(request)获取购物车中的所有商品。

(2) 通过循环语句 for key in cookie_list 遍历购物车中的所有商品,和 3.5.4 节一样,通过语句 response.set_cookie(str(key),1,0)把 cookie 的有效时间设置为 0,从而把商品从购物车中删除。

3. 模板

当所有商品从购物车被删除以后,会返回查看购物车页面,所以,这里的模板与"查看购物车"模块是同一模板。

4. 接口测试

1) 测试用例

表 3-11 为删除购物车中所有商品的测试用例。从购物车中删除所有商品,检验是否购物车中不存在任何商品。

表 3-11 删除购物车中所有商品的测试用例

编号	描述	期望结果
1	把购物车中的所有商品均删除	删除成功,购物车中不存在任何商品

2) XML 数据文件

```
...
<!--把购物车中的所有商品均删除 -->
    <case>
        <TestId>chart-testcase007</TestId>
```

```
        <Title>购物车</Title>
        <Method>get</Method>
        <Desc>把购物车中的所有商品均删除</Desc>
        <Url>http://127.0.0.1:8000/remove_chart_all/</Url>
        <InptArg></InptArg>
        <Result>200</Result>
        <CheckWord>NOT,&lt;td&gt;</CheckWord><!--不存在任何的<td>标识 -->
    </case>
</node>
```

NOT,<td>表示页面中不存在任何<td>标识。

3）测试代码

测试代码不做任何修改。

3.6 送货地址模块

送货地址模块包括"送货地址的添加""送货地址的显示""送货地址的修改"和"送货地址的删除"。

"送货地址的添加"和"送货地址的显示"可以集中在一个文件中进行，用户信息仅显示当前登录用户的地址信息，其他用户的地址信息是不允许显示的。一个用户可以有一到多个送货地址，但是一个送货地址只能对应一个用户，如果两个用户使用的收货地址是相同的，就必须建立地址名称相同的两条记录。

数据模型如下。

```
...
#收货地址
class Address(models.Model):
    user =models.ForeignKey(User)                              #关联用户 id
    address =models.CharField(max_length=50)                   #地址
    phone =models.CharField(max_length=15)                     #电话

    def __str__(self):
        return self.address
...
```

3.6.1 送货地址的添加与显示

1. **urls.py**

```
...
url(r'^add_address/(?P<sign>[0-9]+)/$', views.add_address),
url(r'^view_address/$', views.view_address),
...
```

（1）sign=1 表示从用户信息进入添加送货地址页面。

（2）sign=2 表示从订单信息进入添加送货地址页面。

2. views.py

```
...
#查看地址单信息
def view_address(request):
    util =Util()
    username =util.check_user(request)
    if username=="":
        uf =LoginForm()
        return render(request,"index.html",{'uf':uf,"error":"请登录后再进入"})
    else:
        #返回用户信息
        user_list =get_object_or_404(User, username=username)
        #返回这个用户的地址信息
        address_list =Address.objects.filter(user_id=user_list.id)
        return render(request, 'view_address.html', {"user": username,'addresses': address_list})
...
```

（1）登录用户首先通过语句 user_list = get_object_or_404(User，username=username)获取用户信息。

（2）然后通过语句 address_list = Address.objects.filter(user_id=user_list.id)返回这个登录用户的所有收货地址信息。

（3）最后调用 view_address.html 模板。

```
...
#添加地址
#sign=1 表示从用户信息进入添加送货地址页面
#sign=2 表示从订单信息进入添加送货地址页面
def add_address(request,sign):
    util =Util()
    username =util.check_user(request)
    if username=="":
        uf1 =LoginForm()
        return render(request,"index.html",{'uf':uf1,"error":"请登录后再进入"})
    else:
        #获取当前登录用户的所有信息
        user_list =get_object_or_404(User, username=username)
        #获取当前登录用户的 id
        id =user_list.id
        #判断表单是否提交
        if request.method =="POST":
```

```python
        #如果表单已提交,就准备获取表单信息
        uf =AddressForm(request.POST)
        #验证表单信息是否正确
        if uf.is_valid():
            #如果正确,就开始获取表单信息
            myaddress =(request.POST.get("address", "")).strip()
            phone =(request.POST.get("phone", "")).strip()
            #判断地址是否存在
            check_address =Address.objects.filter(address=myaddress,user_id =id)
            if not check_address:
                #如果不存在,就将表单写入数据库
                address =Address()
                address.address =myaddress
                address.phone =phone
                address.user_id =id
                address.save()
                #返回地址列表页面
                address_list =Address.objects.filter(user_id=user_list.id)
                #如果 sign=="2",就返回订单信息
                if sign=="2":
                    return render(request, 'view_address.html', {"user":
                        username,'addresses': address_list})    #进入订单用户信息
                else:
                    #否则返回用户信息
                    response =HttpResponseRedirect('/user_info/')    #进入用户信息
                    return response
            #否则返回添加用户界面,显示"这个地址已经存在!"的错误信息
            else:
                return render(request,'add_address.html',{'uf':uf,'error':
                    '这个地址已经存在!'})
    #如果没有提交,就显示添加地址页面
    else:
        uf =AddressForm()
    return render(request,'add_address.html',{'uf':uf})
...
```

(1) 登录用户通过语句 user_list = get_object_or_404(User,username=username)获取登录用户信息。

(2) 通过语句 id = user_list.id 获取登录用户数据库中的 id。如果不是送货地址表单提交状态,就显示送货地址表单提交页面,否则通过语句 uf = AddressForm(request. POST)获取表单提交信息。

(3) 获取表单提交信息后,

① 通过判断语句 if uf.is_valid()得知获取数据正确。

② 通过语句 myaddress =（request. POST. get("address"，"")）. strip()和 phone =（request. POST. get("phone"，"")）. strip()获取地址与电话信息。

③ 通过语句 check_address = Address. objects. filter(address=myaddress, user_id=id)与条件语句 if not check_address 判断这个地址在数据库中是否存在。如果已经存在，则调用 add_address. html 模板显示"这个地址已经存在！"的错误信息，否则把地址与电话信息都写入数据库中。

（4）根据 sign=="1"或 sign=="2"返回 user_info()方法或调用 view_address. html 模板。

关于 AddressForm()方法，系统在 forms. py 中进行如下定义。

```python
...
#定义地址表单模型
class AddressForm(forms.Form):
    address = forms.CharField(label='地址', max_length=100)
    phone = forms.CharField(label='电话', max_length=15)
...
```

地址的最大长度定义为 100，电话的最大长度定义为 15，类型都为字符串类型。

3. 模板

添加地址页面 add_address. html。

```html
{%load staticfiles%}
<?Xml version="1.0" encoding="UTF-8"?>
<!DOCTYPE html PUBLIC "-//W3C//DTD XHTML 1.0 Strict//EN""http://www.w3.org/TR/xhtml1/DTD/xhtml1-strict.dtd">
<html xmlns="http://www.w3.org/1999/xhtml" xml:lang="en" lang="en">
<head>
    <meta http-equiv="Content-Type" content="text/html; charset=UTF-8" />
    <title>添加地址</title>
    <!-Bootstrap core CSS ->
    <link href="{%static 'css/signin.css'%}" rel="stylesheet">
    <!-Custom styles for this template ->
    <link href="{%static 'css/bootstrap.min.css'%}" rel="stylesheet">
    <link href="{%static 'css/my.css'%}" rel="stylesheet">
</head>
<body>

    <div class="container">
        <form class="form-signin" method="post" enctype="multipart/form-data">
            <h2 class="form-signin-heading">添加地址</h2>
            {{uf.as_p}}
                <p style="color:red">{{error}}</p><br>
            <button class="btn btn-lg btn-primary btn-block" type="submit">提交
</button><br>
```

```
            </form>

        </div><!--/container -->

</body>
</html>
```

添加送货地址信息如图 3-16 所示。

图 3-16 添加送货地址信息

在生成订单的时候显示地址页面 view_address.html。

```
{%extends "base.html" %}
{%block content %}
                </ul>
                <ul class="nav navbar-nav navbar-right">
                    <li><a href="/user_info/">{{user}}</a></li>
                    <li><a href="/logout/">退出</a></li>
                </ul>
            </div><!--/.nav-collapse -->
        </div>
    </nav>
        <div class="page-header">
            <div id="navbar" class="navbar-collapse collapse">
            </div><!--/.navbar-collapse -->
        </div>
<div class="container theme-showcase" role="main">
<font color="#FF0000">{{error}}</font>
        <div class="row">
            <div class="col-md-6">
            <form method="POST" action="/create_order/">
                <table class="table table-striped">
                    <thead>
                        <tr>
                            <th>选择</th>
                            <th>地址</th>
                            <th>修改</th>
```

```
                    <th>删除</th>
                </tr>
            </thead>
            <tbody>
                {% for address in addresses %}
                <tr>
                    <td><input type="radio" value="{{address.id}}" name="address"></td>
                    <td>{{address.address}}</td>
                    <td><a href="/update_address/{{address.id}}/2/">修改</a></td>
                    <td><a href="/delete_address/{{address.id}}/2/">删除</a></td>
                </tr>
                {% endfor %}
            </tbody>
        </table>
        <input type="submit" value="下一步">
    </form><br>
    <form method="get" action="/add_address/2/">
        <input type="submit" value="添加地址">
    </form>
    </div>
</div>
{% endblock %}
```

显示并且选择用户地址信息如图 3-17 所示。

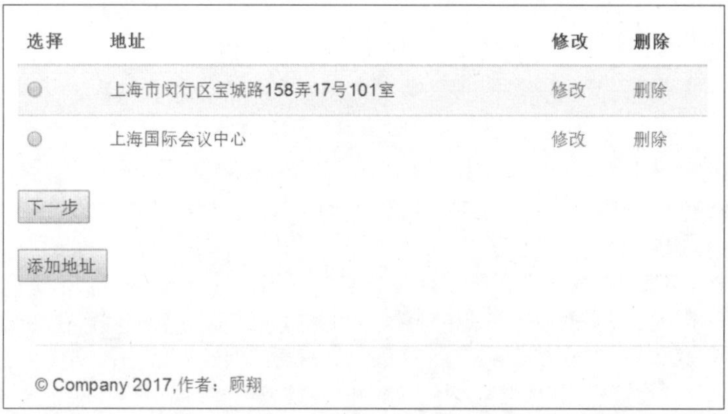

选择	地址	修改	删除
○	上海市闵行区宝城路158弄17号101室	修改	删除
○	上海国际会议中心	修改	删除

下一步

添加地址

© Company 2017,作者：顾翔

图 3-17　显示并且选择用户地址信息

在用户信息中显示地址信息的模板在本书 3.3.3 节中已介绍过，这里不再赘述。

4. 接口测试

1) 测试用例

表 3-12 为送货地址的添加与显示的测试用例。这里设计两个测试用例。

（1）添加一条当前登录用户没有的收货地址信息，系统应该添加成功。

（2）添加一个当前登录用户已经重复的收货地址信息，系统应该报"这个地址已经存在！"的错误信息。

表 3-12 送货地址的添加与显示的测试用例

编号	描述	期望结果
1	为当前登录用户添加一个新的收货地址信息	添加成功，并且可以正确显示
2	为当前登录用户添加一个已经重复的收货地址信息	显示"这个地址已经存在！"的错误信息

2) XML 数据文件

在初始文件 initInfo.xml 中加入地址信息内容。

```xml
...
<!--初始化收货地址信息 -->
    <case>
        <addressid>0</addressid><!--收货地址 id -->
        <address>上海市外滩一号</address><!--收货地址名称 -->
        <phone>13688889999</phone><!--联系电话 -->
        <userid>0</userid><!--用户 id,与初始化用户 id 一致 -->
...
```

地址信息 addressConfig.xml：

```xml
<?xml version="1.0" encoding="UTF-8"?>
<node>
    <case>
        <login>1</login>
    </case>
<!--为当前登录用户添加一个新的地址信息,显示在用户信息中 -->
    <case>
        <TestId>address-testcase001</TestId>
        <Title>地址信息</Title>
        <Method>post</Method>
        <Desc>为当前登录用户添加一个新的地址信息</Desc><!--描述里面有"添加一个新的地址信息",执行操作后程序会把这条记录删除 -->
        <Url>http://127.0.0.1:8000/add_address/1/</Url>
        <InptArg>{"address":"上海市延安中路 100 号","phone":"13681166561"}
</InptArg>
        <Result>200</Result>
        <CheckWord>延安中路</CheckWord><!--与参数中的 address 保持一致 -->
```

```xml
        </case>
        <!--为当前登录用户添加一个已经存在的地址信息 -->
        <case>
            <TestId>address-testcase002</TestId>
            <Title>地址信息</Title>
            <Method>post</Method>
            <Desc>为当前登录用户添加一个已经存在的地址信息</Desc>
            <Url>http://127.0.0.1:8000/add_address/1/</Url>
            <InptArg>{"address":"上海市外滩一号","phone":"13681166561"}</InptArg>
<!--参数中的 address 与初始化地址名称信息保持一致 -->
            <Result>200</Result>
            <CheckWord>这个地址已经存在!</CheckWord>
        </case>
</node>
```

3)测试代码

建立测试代码 addressTest.py,实现方法与前面类似。

```python
#开始测试
def test_address_info(self):
    for mylist in self.mylists:
        data = self.util.run_test(mylist,self.userValues,self.sign)
        #验证返回码
        self.assertEqual(mylist["Result"],str(data.status_code))
        #验证返回文本
        #如果mylist["CheckWord"]标签中存在"NOT"字符串,则调用断言方法 assertNotIn()
        if "NOT" in mylist["CheckWord"]:
            self.assertNotIn(address,str(data.text))
        #否则调用断言方法 assertIn()
        else:
            self.assertIn(mylist["CheckWord"],str(data.text))
        #如果新建立一个地址信息成功,为了保护单个测试用例的独立性,就需要对刚建立的地址信息进行删除操作
        if "添加一个新的地址信息" in mylist["Desc"]:
            payload = eval(mylist["InptArg"])
            address = "\""+(str(payload["address"])).strip()+"\""
            self.dataBase.delete(self.addressTable,"address="+address)
        print(mylist["TestId"]+" is passing!")
```

由于后面"删除收货地址"模块的测试用例中存在对页面中不存在信息的验证,所以事先应书写 if "NOT" in mylist["CheckWord"]验证语句。

3.6.2 送货地址的修改

1. urls.py

```
...
url(r'^update_address/(?P<address_id>[0-9]+)/(?P<sign>[0-9]+)/$',
views.update_address),
...
```

（1）address_id 为修改地址的 id。
（2）sign=1 表示从用户信息进入添加送货地址页面。
（3）sign=2 表示从订单信息进入添加送货地址页面。

2. views.py

```
...
#修改地址信息
#sign=1 表示从用户信息进入添加送货地址页面
#sign=2 表示从订单信息进入添加送货地址页面
def update_address(request,address_id,sign):
    util =Util()
    username =util.check_user(request)
    if username=="":
        uf =LoginForm()
        return render(request,"index.html",{'uf':uf,"error":"请登录后再进入"})
    else:
        #获取指定地址信息
        address_list =get_object_or_404(Address, id=address_id)
        #获取当前登录用户的用户信息
        user_list =get_object_or_404(User, username=username)
        #获取用户 id
        id =user_list.id
        #如果是提交状态
        if request.method =="POST":
            #如果表单已提交,就准备获取表单信息
            uf =AddressForm(request.POST)
            #表单信息验证
            if uf.is_valid():
                #如果数据准确,就获取表单信息
                myaddress =(request.POST.get("address", "")).strip()
                phone =(request.POST.get("phone", "")).strip()
                #判断需要修改地址信息的关联用户是否存在
                check_address =Address.objects.filter(address=myaddress,user_id =id)
                #如果不存在,就修改表单数据并存入数据库
                if not check_address:
```

```
                        Address.objects.filter (id = address_id).update (address =
myaddress,phone =phone)
                        #否则报"这个地址已经存在!"的错误信息
                        else:
                                return render(request,'update_address.html',{'uf':uf,'error':
'这个地址已经存在!','address':address_list})
                        #获取当前登录用户的所有地址信息
                        address_list =Address.objects.filter(user_id=user_list.id)
                        #如果 sign==2,则返回订单信息页面
                        if sign=="2":
                                return render(request, 'view_address.html', {"user": username,
'addresses': address_list})     #进入订单用户信息页面
                        #否则进入用户信息页面
                        else:
                                response =HttpResponseRedirect('/user_info/')     #进入用户
信息页面
                                return response
                        #如果没有提交,就显示修改地址页面
                        else:
                                return render(request,'update_address.html',{'address':address_
list})...
```

（1）登录用户通过语句 address_list = get_object_or_404(Address, id=address_id)获取需要修改的地址信息。

（2）通过语句 user_list = get_object_or_404(User, username=username)获取登录用户信息以及通过语句 id = user_list.id 获取登录用户数据库 id。如果当前不是表单提交状态,则显示修改表单提交信息,否则在验证表单信息合法性后获取表单数据。

（3）接下来验证地址信息在数据库中是否存在,在确保不存在的前提下,通过语句 Address. objects. filter(id=address_id). update(address = myaddress,phone = phone)将其修改内容更新到数据库中。

（4）最后根据 sign==1 或 sign==2 返回 user_info()方法或调用 view_address.html 模板。

3. 模板

update_address.html：

```
{%load staticfiles%}
<?xml version="1.0" encoding="UTF-8"?>
<!DOCTYPE html PUBLIC "-//W3C//DTD XHTML 1.0 Strict//EN" "http://www.w3.org/TR/xhtml1/DTD/xhtml1-strict.dtd">
<html xmlns="http://www.w3.org/1999/xhtml" xml:lang="en" lang="en">
<head>
    <meta http-equiv="Content-Type" content="text/html; charset=UTF-8" />
    <title>修改地址</title>
```

```html
<!--Bootstrap core CSS -->
<link href="{%static 'css/signin.css'%}" rel="stylesheet">
<!--Custom styles for this template -->
<link href="{%static 'css/bootstrap.min.css'%}" rel="stylesheet">
<link href="{%static 'css/my.css'%}" rel="stylesheet">
</head>

<body>

    <div class="container">
        <form class="form-signin" method="post" enctype="multipart/form-data">
            <h2 class="form-signin-heading">修改地址</h2>
            <p><label for="id_address">地址:</label><input type="text" value="{{address.address}}" name="address" id="id_address" size="20" maxlength="100" required /></p>
            <p><label for="id_phone">电话:</label><input type="text" value="{{address.phone}}" name="phone" id="id_phone" size="20" maxlength="15" required /></p>
            <p style="color:red">{{error}}</p><br>
            <button class="btn btn-lg btn-primary btn-block" type="submit">修改</button><br>
        </form>

    </div><!--/container -->

</body>
</html>
```

由于在修改的时候需要显示以前的地址信息内容,所以不能使用 AddressForm 类,只能用 HTML 把 form 信息写出来,如图 3-18。

图 3-18　修改收货地址

4.接口测试

1) 测试用例

表 3-13 为送货地址修改的测试用例。这里设计两个测试用例。

(1) 修改一个当前登录用户没有添加过的收货地址信息,系统应该修改成功。

(2) 修改一个当前登录用户已经添加过的地址信息,系统应该报"这个地址已经存在!"的错误信息。

表 3-13 送货地址修改的测试用例

编号	描 述	期望结果
1	修改一条当前登录用户没有添加过的收货地址信息	修改成功,并且可以正确显示
2	修改一条当前登录用户已经添加过的收货地址信息	显示"这个地址已经存在!"的错误信息

2) XML 数据文件

在 addressConfig.xml 文件中加入如下内容。

```xml
...
<!--修改一条当前登录用户没有使用过的收货地址信息 -->
<case>
    <TestId>address-testcase003</TestId>
    <Title>地址信息</Title>
    <Method>post</Method>
    <Desc>修改一条当前登录用户没有添加过的收货地址信息</Desc>
    <Url>http://127.0.0.1:8000/update_address/0/1/</Url>
    <InptArg>{"address":"上海市延安中路 100 号","phone":"13681166561"}</InptArg>
    <Result>200</Result>
    <CheckWord>上海市延安中路 100 号</CheckWord><!--与参数中的 phone 保持一致 -->
</case>
<!--修改一条当前登录用户已经添加过的收货地址信息 -->
<case>
    <TestId>address-testcase004</TestId>
    <Title>地址信息</Title>
    <Method>post</Method>
    <Desc>修改一条当前登录用户已经使用过的收货地址信息</Desc>
    <Url>http://127.0.0.1:8000/update_address/0/1/</Url>
    <InptArg>{"address":"上海市延安中路 100 号","phone":"13681166561"}</InptArg>
    <Result>200</Result>
    <CheckWord>这个地址已经存在!</CheckWord><!--与测试用例 address-testcase003 参数中的 address 保持一致 -->
</case>
...
```

测试用例 address-testcase004 是基于 address-testcase003 基础上的,这里把 address-testcase003 `<InptArg>…</InptArg>` 参数信息原封不变地进行了复制。

3) 测试代码

测试代码保持不变。

3.6.3 送货地址的删除

1. urls.py

```
...
url(r'^delete_address/(?P<address_id>[0-9]+)/(?P<sign>[0-9]+)/$',
views.delete_address),
...
```

(1) address_id 为删除地址的 id。
(2) sign=1 表示从用户信息进入删除送货地址页面。
(3) sign=2 表示从订单信息进入删除送货地址页面。

2. views.py

```python
...
#删除地址信息
#sign=1 表示从用户信息进入删除送货地址页面
#sign=2 表示从订单用户信息进入删除送货地址页面
def delete_address(request,address_id,sign):
    util =Util()
    username =util.check_user(request)
    if username=="":
        uf =LoginForm()
        return render(request,"index.html",{'uf':uf,"error":"请登录后再进入"})
    else:
        #获取指定地址信息
        user_list =get_object_or_404(User, username=username)
        #删除这个地址信息
        Address.objects.filter(id=address_id).delete()
        #返回地址列表页面
        address_list =Address.objects.filter(user_id=user_list.id)
        #如果 sign==2,就返回订单信息页面
        if sign=="2":
            return render(request, 'view_address.html', {"user": username,'addresses':address_list})      #进入订单用户信息页面
        #否则进入用户信息页面
        else:
            response =HttpResponseRedirect('/user_info/')     #进入用户信息页面
            return response
...
```

(1) 登录用户通过语句 user_list = get_object_or_404(User，username=username)获得用户信息。
(2) 通过语句 Address.objects.filter(id=address_id).delete()删除需要删除的收货地

址信息。

(3) 通过语句 address_list = Address.objects.filter(user_id=user_list.id) 返回当前用户的所有收货地址信息。

(4) 最后根据 sign=="1" 或 sign=="2" 调用 user_info() 方法，或调用 view_address.html 模板。

3. 模板

删除送货地址以后，根据参数 sign 进入用户信息页面或者生成订单选择地址页面。所以，这里没有专门的模板文件。

4. 接口测试

1) 测试用例

表 3-14 为删除送货地址的测试用例。删除当前登录用户的一个送货地址信息，系统应该删除成功，在收货地址显示页面中不显示这条记录。

表 3-14　删除送货地址的测试用例

编号	描　　述	期望结果
1	删除当前登录用户的一个送货地址信息	删除成功，在收货地址显示页面中不显示这条记录

使用测试程序往数据库中插入用户地址信息，然后测试产品代码是否可以正确地进行删除操作。

2) XML 数据文件

在 addressConfig.xml 文件中加入如下内容。

```
...
    <!--删除地址信息 -->
    <case>
        <TestId>address-testcase005</TestId>
        <Title>地址信息</Title>
        <Method>get</Method>
        <Desc>删除地址信息</Desc>
        <Url>http://127.0.0.1:8000/delete_address/1234/1/</Url><!-- "1234"作为测试程序地址 id 插入数据库表中 -->
        <InptArg></InptArg>
        <Result>200</Result>
        <CheckWord>NOT,朝阳门外 5454 号</CheckWord><!--NOT 后的字符将作为地址名称插入数据库中，验证这个信息删除后不能被显示-->
    </case>
</node>
```

标签 <Url>…</Url> 中的 "1234" 作为测试程序地址 id 插入数据库表中，标签 <CheckWord>…</CheckWord> 中 "NOT," 后的字符串 "朝阳门外 5454 号" 将作为地址名称插入数据库中，验证这个信息删除后不能被显示。所以，这里不管是地址 id "1234"，还是地址信息 "朝阳门外 5454 号"，都可以任意修改。在此特别说明，为了保证每个测试用例的独立性,每个测试用例尽可能做到不依赖其他测试用例。如果需要依赖,必须把依赖的测试用例联合在一起执行。

3) 测试代码

在测试代码中,在运行测试的开始做了一个判断。

```
...
    #开始测试
    def test_address_info(self):
        for mylist in self.mylists:
            if ("NOT" in mylist["CheckWord"]):
                #id 从 mylist["Url"]中获取
                id =(mylist["Url"]).split("/")[4]
                #address 从 mylist["CheckWord"]中获取
                address= (mylist["CheckWord"]).split(",")[1]
                addressvalues =id+",'"+address+"','13666666666',0"
                self.util.insertTable(self.dataBase,self.addressTable, addressvalues)
...
```

CheckWord 里存在"NOT",表示这条记录为需要删除地址信息的测试用例。在删除这条记录前,需要先建立这条记录。

(1) 通过配置文件中的(mylist["Url"]).split("/")[4]获取需要建立收货地址记录的 id。

(2) 通过(mylist["CheckWord"]).split(",")[1]获取这条收货地址记录的地址信息。

(3) 电话号码硬编码为"13666666666"。

需要特别指出的是,这里的代码不删除初始化信息,是为了尽可能保证每条测试用例之间的相互独立性。

在 3.5.4 节中,为了保持每条测试用例的独立性,也应该单独在购物车中建立一个商品,然后删除这个新建立的商品,而不是删除在初始化中建立的购物车中的商品,读者可以自己去完善修改。

3.7 订单模块

订单模块包括"总订单的生成和显示""查看所有订单""删除单个订单"以及"删除总订单"。由于一个总订单关联多个订单,并且订单与用户、商品以及用户收货地址都有相应的对应关系,所以这里程序处理的业务逻辑是比较复杂的。单个订单的数据模型如下。

```
...
#单个订单
class Order(models.Model):
    order =models.ForeignKey(Orders)           #关联总订单 id
    user =models.ForeignKey(User)              #关联用户 id
```

```
    goods = models.ForeignKey(Goods)                        #关联商品id
    count = models.IntegerField()                           #数量
...
```

总订单的数据模型如下。

```
...
#总订单
class Orders(models.Model):
    address = models.ForeignKey(Address)                    #关联送货地址id
    create_time = models.DateTimeField(auto_now=True)       #创建时间
    status = models.BooleanField()                          #订单状态

    def __str__(self):
        return self.create_time
...
```

3.7.1 总订单的生成和显示

1. urls.py

```
...
url(r'^create_order/$', views.create_order),
url(r'^view_order/(?P<order_id>[0-9]+)/$', views.view_order),
...
```

orders_id 为单个订单的 id。

要显示一个订单首先是生成一个订单,然后把这个生成的订单显示出来。

2. views.py

create_order()方法用于生成订单。

```
...
#生成订单信息
def create_order(request):
    util = Util()
    username = util.check_user(request)
    if username == "":
        uf = LoginForm()
        return render(request,"index.html",{'uf':uf,"error":"请登录后再进入"})
    else:
        #根据登录的用户名获得用户信息
        user_list = get_object_or_404(User, username=username)
        #从选择地址信息中获得建立这个订单的送货地址id
        address_id = (request.POST.get("address", "")).strip()
        #如果没有选择地址,就返回错误提示信息
```

```python
        if address_id=="":
            address_list =Address.objects.filter(user_id=user_list.id)
            return render(request, 'view_address.html', {"user": username,'addresses':
address_list,"error":"必须选择一个地址!"})
        #否则开始形成订单
        else:
            #把数据存入数据库中的总订单表中
            orders =Orders()
            #获得订单的送货地址 id
            orders.address_id =int(address_id)
            #设置订单的状态为未付款
            orders.status =False
            #保存总订单信息
            orders.save()
            #准备把订单中的每个商品存入单个订单表
            #获得总订单 id
            orders_id =orders.id
            #获得购物车中的内容
            cookie_list =util.deal_cookies(request)
            #遍历购物车
            for key in cookie_list:
                #构建对象 Order()
                order =Order()
                #获得总订单 id
                order.order_id =orders_id
                #获得用户 id
                order.user_id =user_list.id
                #获得商品 id
                order.goods_id =key
                #获得数量
                order.count =int(cookie_list[key])
                #保存单个订单信息
                order.save()
            #清除所有 cookies,并且显示这个订单
            response =HttpResponseRedirect('/view_order/'+str(orders_id))
            for key in cookie_list:
                response.set_cookie(str(key),1,0)
            return response
...
```

（1）登录用户通过语句 user_list = get_object_or_404(User, username=username)显示登录用户信息。

（2）通过语句 address_id =（request.POST.get("address", "")）.strip()获取订单对应的收货地址。

(3) 如果没有选择,就调用 view_address.html 模板,显示"必须选择一个地址!"的错误信息。

(4) 否则开始生成订单,通过语句 orders = Orders()获得总订单类变量 orders。

① 通过语句 orders.address_id = int(address_id)和 orders.status = False 设置订单的地址 id 以及字符状态为 False,即"未支付"。

② 通过语句 orders.save()把总订单信息保存在数据库中。

③ 存储完毕,系统生成总订单中的每个商品订单信息,通过语句 orders_id = orders.id 获得总订单 id。

④ 通过语句 cookie_list = util.deal_cookies(request)获得购物车中的所有商品。

⑤ 通过循环语句 for key in cookie_list 遍历这些商品。

⑥ 通过语句 order = Order()建立单个订单类变量 order。

⑦ 通过语句 order.order_id = orders_id、order.user_id = user_list.id、order.goods_id = key 和 order.count = int(cookie_list[key])向 order 类变量赋予总订单 id、用户 id、商品 id 以及商品数量。

⑧ 通过语句 order.save()将订单信息保存到数据库中。

⑨ 通过循环语句 for key in cookie_list 遍历购物车里的所有商品。

⑩ 最后利用语句 response.set_cookie(str(key),1,0)清除购物车中的所有商品。

方法 view_order()用于显示单个订单信息。

```
    ...
    #显示订单
    def view_order(request,orders_id):
        util =Util()
        username =util.check_user(request)
        if username=="":
            uf =LoginForm()
            return render(request,"index.html",{'uf':uf,"error":"请登录后再进入"})
        else:
            #获取总订单信息
            orders_filter =get_object_or_404(Orders,id=orders_id)
            #获取订单的收货地址信息
            address_list =get_object_or_404(Address,id=orders_filter.address_id)
            #获取收货地址信息中的地址
            address =address_list.address
            #获取单个订单表中的信息
            order_filter =Order.objects.filter(order_id=orders_filter.id)
            #建立列表变量 order_list,里面存放的是每个 Order_list 对象
            order_list_var =[]
            prices=0
            for key in order_filter:
                #定义 Order_list 对象
                order_object =Order_list
```

```
            #产生一个 Order_list 对象
            order_object =util.set_order_list(key)
            #把当前 Order_list 对象加入到列表变量 order_list
            order_list_var.append(order_object)
            #获取当前商品的总价格
            prices =order_object.price * order_object.count +prices
        return render(request, 'view_order.html', {"user": username,'orders':
orders_filter,'order': order_list_var,'address': address," prices": str
(prices)})
        ...
```

(1) 登录用户通过语句 orders_filter = get_object_or_404(Orders,id=orders_id)获得总订单信息。

(2) 通过语句 address_list = get_object_or_404(Address,id=orders_filter.address_id)获得总订单的收货地址信息以及通过语句 address = address_list.address 获得收货地址信息中的地址内容信息。

(3) 通过语句 order_filter = Order.objects.filter(order_id=orders_filter.id)返回所有该订单下的单个订单信息。

(4) 通过循环语句 for key in order_filter 遍历所有的单个订单信息。

(5) 在循环体内通过 order_object = Order_list 语句定义 Order_list 对象。

(6) 通过 order_object =util.set_order_list(key)语句调用 Util 类中的 set_order_list 方法(下面介绍),返回 Order_list 对象。

(7) 由语句 order_list_var.append(order_object)把 Order_list 对象封装在 order_list_var 变量中(order_list_var 变量在循环外被初始化)。

(8) 由语句 prices = order_object.price * order_object.count + prices 计算总订单中所有的商品价格。

(9) 最后调用 view_order.html 模板。

调用模板的参数中包括

① user:用户名。

② orders:总订单信息。

③ order:单个订单列表信息,里面是多个 Order_list 对象。

④ address:收货地址信息。

⑤ prices:总价格信息。

方法 set_order_list()在 goods/util.py 中定义。

```
        ...
        #定义单个订单变量
        def set_order_list(self,key):
            order_list =Order_list()
            order_list.set_id(key.id)                                    #主键
            good_list =get_object_or_404(Goods,id=key.goods_id)          #获得当前商品信息
```

```
            order_list.set_good_id(good_list.id)      #订单中的商品 id
            order_list.set_name(good_list.name)       #订单中的商品名称
            order_list.set_price(good_list.price)     #订单中的商品价格
            order_list.set_count(key.count)           #购买数量
            return order_list
    ...
```

方法返回的是 Order_list 类。Order_list 类在 object.py 中定义。

```
...
#订单模型
class Order_list():
    #订单 id
    def set_id(self,id):
        self.id=id

    #订单中的商品 id
    def set_good_id(self,good_id):
        self.good_id=good_id

    #订单中的商品名称
    def set_name(self,name):
        self.name=name

    #订单中的商品价格
    def set_price(self,price):
        self.price=price

    #订单中的商品数量
    def set_count(self,count):
        self.count=count

    #商品总价格(个数×单价)
    def set_prices(self,prices):
        self.prices=prices
...
```

3. 模板

```
{%extends "base.html"%}
{%block content%}
            </ul>
            <ul class="nav navbar-nav navbar-right">
                <li><a href="/user_info/">{{user}}</a></li>
```

```html
            <li><a href="/logout/">退出</a></li>
          </ul>
        </div><!--/.nav-collapse -->
      </div>
    </nav>

    <div class="container theme-showcase" role="main">

      <div class="page-header">
        <div id="navbar" class="navbar-collapse collapse">
        </div><!--/.navbar-collapse -->
      </div>
      <p>生成时间:{{orders.create_time}} 配货地址:{{address}}总价格：¥ {{prices}}</p>

      <div class="row">
        <div class="col-md-6">
          <table class="table table-striped">
            <thead>
              <tr>
                <th>编号</th>
                <th>商品名称</th>
                <th>商品价格</th>
                <th>个数</th>
                <th>删除</th>
              </tr>
            </thead>
            <tbody>
              {% for key in order %}
                <tr>
                  <td><a href="/view_goods/{{key.good_id}}/">{{key.id}}</a></td>
                  <td>{{ key.name }}</td>
                  <td> ¥ {{ key.price }}</td>
                  <td>{{ key.count }}</td>
                  <td><a href="/delete_orders/{{key.id}}/3/">删除</a></td>
                </tr>
              {% endfor %}
            </tbody>
          </table>
          <input type="submit" value="支付">
        </div>
```

```
            </div>
{%endblock%}
```

总订单信息通过<p>生成时间：{{orders.create_time}}配货地址：{{address}}总价格：￥{{prices}}</p>显示订单信息，通过{% for key in order %}遍历显示，如图 3-19 所示。

图 3-19　显示当前生成的订单

4. 接口测试

1）测试用例

表 3-15 为生成一个订单的测试用例。测试程序通过初始化一个订单数据和对应的总订单数据，生成一个订单和对应的总订单，最后验证生成的订单是否可以正确地被显示。

表 3-15　生成一个订单的测试用例

编号	描　　述	期望结果
1	生成并且显示当前用户的一个订单	生成订单并且正确显示

2）XML 数据文件

首先在 initInfo.xml 中建立订单信息。

```
<!--初始化总订单信息 -->
    <case>
        <ordersid>0</ordersid><!--总订单 id -->
        <createtime>Sept. 8, 2017, 6:29 a.m.</createtime><!--总订单产生日期 -->
        <status>1</status><!--这里必须设置为 1,表示已支付,与测试代码区别 -->
        <ordersaddressid>0</ordersaddressid><!--地址 id,与初始化地址 id 保持一致
-->
    </case>
<!--初始化单个订单信息 -->
    <case>
        <orderid>0</orderid><!--单个订单 id -->
        <count>9999</count><!--单个订单数量,这里必须填 9999,与测试代码区别-->
        <ordergoodid>0</ordergoodid><!--商品 id,与初始化商品 id 保持一致 -->
        <orderorderid>0</orderorderid><!--总订单 id,与初始化总订单 id 保持一致 -->
```

```
            <orderuserid>0</orderuserid><!--用户id,与初始化用户id保持一致 -->
        </case>
</node>
```

由于在这里要使用到cookie,所以在测试程序开始要通过程序代码向购物车中添加一个商品。这里的测试用例与购物车的测试用例第一条一样。建立测试配置文件orderConfig.xml。

```xml
<?xml version="1.0" encoding="UTF-8"?>
<node>
    <case>
        <login>1</login>
    </case>
    <!--添加商品到购物车,查看购物车内商品数量的变化 -->
    <case>
        <TestId>order-testcase001</TestId>
        <Title>购物车</Title>
        <Method>get</Method>
        <Desc>添加到购物车</Desc>
        <Url>http://127.0.0.1:8000/add_chart/0/1/</Url>
        <InptArg></InptArg>
        <Result>200</Result>
        <CheckWord>查看购物车 &lt;font color=&quot;#FF0000&quot;&gt;1&lt;/font&gt;&lt;/a&gt;</CheckWord><!--显示添加成功 -->
    </case>
    <!--生成并且显示当前用户的一个订单 -->
    <case>
        <TestId>order-testcase002</TestId>
        <Title>订单信息</Title>
        <Method>post</Method>
        <Desc>生成并且显示当前用户的一个订单</Desc>
        <Url>http://127.0.0.1:8000/create_order/</Url>
        <InptArg>{"address":"0"}</InptArg>
        <Result>200</Result>
        <CheckWord>1234.56</CheckWord><!--检查单个订单中的价格信息是否正确显示,order-testcase002必须为建立订单 -->
    </case>
```

3) 测试代码

orderTest.py:

```python
#!/usr/bin/env python
#coding:utf-8
import unittest,requests
from util import GetXML,DB,Util
```

```python
class orderTest(unittest.TestCase):
    def setUp(self):
        print("--------测试开始--------")
        xmlfile = "orderConfig.xml"
        #建立 GetXML 对象变量
        xmlInfo = GetXML()
        #获取是否需要登录的信息
        self.sign = xmlInfo.getIsLogin(xmlfile)
        #获取所有测试数据
        self.mylists = xmlInfo.getxmldata(xmlfile)
        #建立 DB 变量
        self.dataBase = DB()
        #建立 util 变量
        self.util = Util()
        #初始化用户记录
        self.userTable = "goods_user"
        self.userValues = self.util.inivalue(self.dataBase, self.userTable, "0")
        #初始化商品记录
        self.goodTable = "goods_goods"
        self.goodValues = self.util.inivalue(self.dataBase, self.goodTable, "1")
        #初始化地址记录
        self.addressTable = "goods_address"
        self.addressValues = self.util.inivalue(self.dataBase, self.addressTable, "2")
        #初始化总订单记录
        self.ordersTable = "goods_orders"
        self.ordersValues = self.util.inivalue(self.dataBase, self.ordersTable, "3")
        #初始化订单记录
        self.orderTable = "goods_order"
        self.orderValues = self.util.inivalue(self.dataBase, self.orderTable, "4")
    #开始测试
    def test_order_info(self):
        for mylist in self.mylists:
            data = self.util.run_test(mylist, self.userValues, self.sign)
            #验证返回码
            self.assertEqual(mylist["Result"], str(data.status_code))
            #验证返回文本
            #如果 mylist["CheckWord"]标签中存在"NOT"字符串,则调用断言方法 assertNotIn()
            if "NOT" in mylist["CheckWord"]:
                self.assertNotIn((mylist["CheckWord"]).split(",")[1], str(data.text))
```

```python
            #建立单独订单记录
            self.util.insertTable(self.dataBase,self.ordersTable,self.ordersValues)
            #建立总订单记录
            self.util.insertTable(self.dataBase,self.orderTable,self.orderValues)
            #否则调用断言方法 assertIn()
        else:
                self.assertIn(mylist["CheckWord"],str(data.text))
            #如果测试用例的目的是查看所有订单,测试完毕就需要删除测试数据
            if "view_all_order" in mylist["Url"]:
                self.dataBase.delete(self.ordersTable,"status='0'")
                self.dataBase.delete(self.orderTable,"count=1")
            print (mylist["TestId"]+" is passing!")
    def tearDown(self):
            #删除 setup 建立的单个订单
            self.util.tearDown(self.dataBase,self.orderTable,self.orderValues)
            #删除 setup 建立的总订单
            self.util.tearDown(self.dataBase,self.ordersTable,self.ordersValues)
            #删除 setup 建立的地址
            self.util.tearDown(self.dataBase,self.addressTable,self.addressValues)
            #删除 setup 建立的商品
            self.util.tearDown(self.dataBase,self.goodTable,self.goodValues)
            #删除 setup 建立的用户
            self.util.tearDown(self.dataBase,self.userTable,self.userValues)
            #关闭数据库连接
            self.dataBase.close()
            print("--------测试结束--------")
if __name__=='__main__':
        #构造测试集
        suite=unittest.TestSuite()
        suite.addTest(orderTest("test_order_info"))
        #运行测试集合
        runner=unittest.TextTestRunner()
        runner.run(suite)
```

由于这里要用到用户、商品、收货地址、单个订单和总订单信息,所以在 Setup()方法中要对它们进行初始化。由于测试用例"显示当前用户的所有订单"必须用到测试用例"生成并且显示当前用户的一个订单"建立的测试记录,所以在测试用例"显示当前用户的所有订单"后删除测试用例"生成并且显示当前用户的一个订单"建立的测试数据(参见代码中的注释"#如果测试用例的目的是查看所有订单,测试完毕就需要删除测试数据")(这里没有把两个测试用例相互独立出来,读者可以考虑如何将这两个测试用例分开,使得它们互相独立)。

3.7.2 查看所有订单

1. urls.py

```
...
url(r'^view_all_order/$', views.view_all_order),
...
```

2. views.py

```
...
#查看所有订单
def view_all_order(request):
    util = Util()
    username = util.check_user(request)
    if username == "":
        uf = LoginForm()
        return render(request,"index.html",{'uf':uf,"error":"请登录后再进入"})
    else:
        #获得所有总订单信息
        orders_all = Orders.objects.all()
        #初始化订单结果列表,这个列表变量在本段代码最后传递给模板文件
        Reust_Order_list = []
        #遍历总订单
        for key1 in orders_all:
            #通过当前订单 id 获取这个订单的单个订单详细信息
            order_all = Order.objects.filter(order_id=key1.id)
            #检查这个订单是否属于当前用户
            user = get_object_or_404(User,id=order_all[0].user_id)
            #如果属于当前用户,就将其放入总订单列表中
            if user.username == username:
                #初始化总订单列表
                Orders_object_list = []
                #初始化总订单类
                orders_object = Orders_list
                #产生一个 Orders_list 对象
                orders_object = util.set_orders_list(key1)
                #初始化总价格为 0
                prices = 0
                #遍历这个订单
                for key in order_all:
                    #初始化订单类
                    order_object = Order_list
                    #产生一个 Order_list 对象
                    order_object = util.set_order_list(key)
```

```
            #将产生的 order_object 类加到总订单列表中
            Orders_object_list.append(order_object)
            #计算总价格
            prices = order_object.price * key.count + prices
    #把总价格放到 order_object 类中
    order_object.set_prices(prices)
    #把当前记录加到 Reust_Order_list 列中
    #从这里可以看出,Reust_Order_list 中的每一项都是一个字典类型,key 为
    #总订单类 orders_object,value 为总订单列表 Orders_object_list
    #总订单列表 Orders_object_list 中的每一项都为一个单独订单对象
    #order_object,即 Reust_Order_list=[{orders_object 类:[order_object
    #类,…]},…]
    Reust_Order_list.append({orders_object:Orders_object_list})
    return render(request, 'view_all_order.html', {"user": username,'Orders_set': Reust_Order_list})
    …
```

(1) 登录用户通过语句 orders_all = Orders.objects.all()获得数据库中的所有总订单信息。

(2) 通过循环语句 for key1 in orders_all 遍历总订单,在循环体内通过语句 order_all = Order.objects.filter(order_id=key1.id)获得当前总订单下的所有单个订单。

(3) 通过语句 user = get_object_or_404(User,id=order_all[0].user_id)获得单个订单的用户信息,由判断语句 if user.username == username 判断这个订单是否属于当前登录用户,只有属于当前登录用户的订单信息才可以被显示出来。

(4) 通过语句 orders_object = Orders_list 初始化一个总订单类对象。

(5) 通过语句 orders_object = util.set_orders_list(key1)调用 Util 类中的 set_orders_list()方法获得总订单类对象。

(6) 通过语句 prices=0 初始化总价格为 0,由循环语句 for key in order_all 遍历当前总订单下的所有单个订单。

(7) 在循环体内由语句 order_object = Order_list 和 order_object = util.set_order_list(key)初始化并且获得单个订单类 order_object 对象。

(8) 再由语句 Orders_object_list.append(order_object)把单个订单类 order_object 对象加到 Orders_object_list 列表变量中,这里的 Orders_object_list 列表变量是在第一个循环后和第二个循环前被初始化的。

(9) 通过语句 prices = order_object.price * key.count + prices 累积计算这个总订单内商品的总价格,第二个循环结束,继续第一个循环。

(10) 通过语句 order_object.set_prices(prices)把总价格加到 order_object 类中。

(11) 把 orders_object 和 Orders_object_list 以值参对的形式加到 Reust_Order_list 列表变量中。

(12) 通过语句 Reust_Order_list.append({orders_object:Orders_object_list})把参数 {orders_object:Orders_object_list} 加到列表变量 Reust_Order_list 后,这里的列表变量

Reust_Order_list 是在第一个循环前初始化的。

（13）最后调用 view_all_order.html 模板。

这里调用模板的变量 Reust_Order_list 是一个比较复杂的数据结构，首先它是一个列，每个类中包含一个字典类型，这个字典类型的参数为总订单类 orders_object，值为总订单列表 Orders_object_list。总订单列表 Orders_object_list 中的每一项为一个单独订单对象 order_object，可以标记为 Reust_Order_list＝[{orders_object 类：[order_object 类，…]}，…]。

set_orders_list()方法在 goods/util.py 中定义。

```python
...
def set_orders_list(self,key):
    order_list =Orders_list()
    order_list.set_id(key.id)                          #主键
    order_list.set_address(key.address)                #地址信息
    order_list.set_create_time(key.create_time)        #创建时间
    return order_list
...
```

set_orders_list()方法返回 Orders_list 对象。Orders_list 对象在 object.py 中定义。

```python
...
#总订单模型
class Orders_list():
    #总订单 id
    def set_id(self,id):
        self.id=id
    #总订单收货地址
    def set_address(self,address):
        self.address=address
    #总订单建立时间
    def set_create_time(self,create_time):
        self.create_time=create_time
...
```

3. 模板

view_all_order.html：

```html
{%extends "base.html" %}
{%block content %}

            </ul>
                <ul class="nav navbar-nav navbar-right">
                    <li><a href="/user_info/">{{user}}</a></li>
                    <li><a href="/logout/">退出</a></li>
```

```html
            </ul>
        </div><!--/.nav-collapse -->
    </div>
</nav>

<div class="container theme-showcase" role="main">

    <div class="page-header">
        <div id="navbar" class="navbar-collapse collapse">
        </div><!--/.navbar-collapse -->
    </div>
    {% for key1 in Orders_set %}
    <div class="row">
        <div class="col-md-6">
            <table class="table table-striped">
                <thead>
                    <tr>
                        <th>编号</th>
                        <th>商品名称</th>
                        <th>商品价格</th>
                        <th>个数</th>
                        <th>删除</th>
                    </tr>
                </thead>
                <tbody>
                    {% for key2,value in key1.items %}
                    {% for key in value %}
                        <tr>
                            <td><a href="/view_goods/{{key.good_id}}/">{{ key.id }}</a></td>
                            <td>{{ key.name }}</td>
                            <td> ¥ {{ key.price }}</td>
                            <td>{{ key.count }}</td>
                            <td><a href="/delete_orders/{{key.id}}/1/">删除</a></td>
                        </tr>
                    {% endfor %}
                    订单编号:{{key2.id}},创建时间:{{key2.create_time}},地址:{{key2.address}}<br>
                    {% if not key2.status %}
                        <input type="submit" value="支付">
                    {% endif %}
                        <a href="/delete_orders/{{key2.id}}/2/">删除</a>
                    {% endfor %}
```

```
                    </tbody>
                </table>
            </div>
        {% endfor %}

    </div>
{% endblock %}
```

(1) 通过{% for key1 in Orders_set %}遍历Orders_set中的每个字典类型。

(2) 通过{% for key2,value in key1.items %}遍历字典类型中的每个参数和值。

(3) 通过{% for key in value %}遍历key值下的每个单独订单类型。

(4) 通过{% if not key2.status %}判断当前订单是否支付，如果没有支付，就显示"支付"按钮，如图3-20所示。

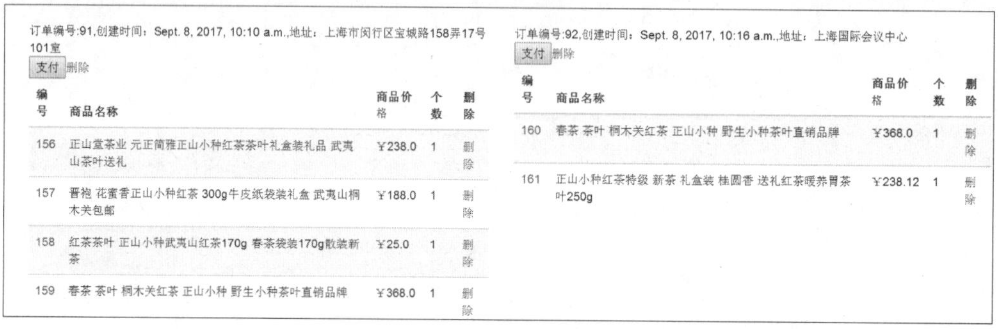

图3-20 查看所有订单

4. 接口测试

1) 测试用例

表3-16为生成所有订单的测试用例。与测试单个订单一样，测试程序通过初始化一个订单数据和对应的总订单数据，生成一个订单和对应的总订单，最后验证这些订单可以被正确显示。

表3-16 生成所有订单的测试用例

编号	描述	期望结果
1	显示当前用户的所有订单	当前用户的所有订单被正确显示

2) XML数据文件

"显示当前用户的所有订单"测试用例的测试数据在orderConfig.xml中定义如下。

```
...
<!--显示当前用户的所有订单 -->
    <case>
        <TestId>order-testcase003</TestId>
        <Title>订单信息</Title>
```

```
            <Method>get</Method>
            <Desc>显示当前用户的所有订单</Desc>
            <Url>http://127.0.0.1:8000/view_all_order/</Url>
            <InptArg></InptArg>
            <Result>200</Result>
            <CheckWord>上海市外滩一号</CheckWord><!--检查订单中的地址信息是否被正确
显示 -->
    </case>
...
```

3) 测试代码

测试代码保持不变。

3.7.3 删除订单

1. urls.py

```
...
url(r'^delete_orders/(?P<orders_id>[0-9]+)/(?P<sign>[0-9]+)/$',views.delete
_orders),
...
```

（1）orders_id 为删除的单个订单或者总订单 id。
（2）sign=1 或者 3 表示删除单个订单。
（3）sign=2 表示删除总订单。
（4）sign=1 或者 2 表示从查看总订单进入删除订单页面。
（5）sign=3 表示从查看单个订单进入删除订单页面。

2. views.py

```
...
#删除订单
#id=1,3删除单个订单,id=2删除总订单
#id=1,2从查看总订单进入,id=3从查看单个订单进入
def delete_orders(request,orders_id,sign):
    util=Util()
    username=util.check_user(request)
    if username=="":
        uf=LoginForm()
        return render(request,"index.html",{'uf':uf,"error":"请登录后再进入"})
    else:
        #如果删除单个订单
        if sign=="1" or sign=="3":
            #通过主键获得单个订单的内容
            order_filter=get_object_or_404(Order,id=orders_id)
```

```
            #获取当前订单所属的总订单
            orders_filter =get_object_or_404(Orders,id=order_filter.order_id)
            #删除这个单个订单
            Order.objects.filter(id=orders_id).delete()
            #判断这个总订单下是否还有商品
            judge_order =Order.objects.filter(order_id =orders_filter.id)
            #如果没有商品了
            if (len(judge_order))==0:
                #就删除这个订单所处的总订单记录
                Orders.objects.filter(id=orders_filter.id).delete()
                #如果标记为 3,则调用 goods_view()方法
                if sign=="3":
                    response =HttpResponseRedirect('/goods_view/')
                    #调用 goods_view()方法
                #如果标记为 1,则调用 view_all_order()方法
                if sign=="1":
                    response =HttpResponseRedirect('/view_all_order/')
                    #调用 view_all_order()方法
            #如果还有商品,且标记为 3,则调用 view_order()方法
            elif sign=="3":
                response = HttpResponseRedirect ('/view_order/' + str (orders_
                filter.id)+'/') #调用 view_order()方法
            #否则调用 view_all_order()方法
            else:
                response =HttpResponseRedirect('/view_all_order/')
                #调用 view_all_order()方法
        #如果删除总订单
        if sign =="2":
            #删除单个订单
            Order.objects.filter(order_id=orders_id).delete()
            #删除总订单
            Orders.objects.filter(id=orders_id).delete()
            #返回查看所有订单页面
            response =HttpResponseRedirect('/view_all_order/')
            #调用 view_all_order()方法
        return response
    ...
```

(1) 登录用户通过判断语句 if sign == "1" or sign=="3"决定是否删除单独一个订单。

(2) 如果是删除单独一个订单,

① 通过语句 order_filter = get_object_or_404(Order,id=orders_id)获得单独订单信息。

② 通过语句 orders_filter = get_object_or_404(Orders,id=order_filter.order_id)获

得当前订单所属的总订单信息。

③ 通过语句 Order.objects.filter(id=orders_id).delete()删除这个订单。注意,如果这个订单所属的总订单都没有订单了,就必须删除这个订单。

④ 通过语句 judge_order = Order.objects.filter(order_id = orders_filter.id)判断这个订单所属的总订单是否没有订单了,如果判断语句 if (len(judge_order))==0 的结果为真,那么通过语句 Orders.objects.filter(id=orders_filter.id).delete()将这条订单所属的总订单删除,然后根据 sign=="1"或 sign=="3"调用 goods_view()方法或 view_order()方法。

(3) 如果 if sign=="2",也就是说删除的是总订单,那么先通过语句 Order.objects.filter(order_id=orders_id).delete()删除这个总订单下的所有单个订单,然后通过语句 Orders.objects.filter(id=orders_id).delete()删除这个总订单,最后调用方法 view_all_order()。

(4) 删除单个订单可以从订单确认页面进入,也可以从查看所有订单页面进入。删除总订单只能从查看所有订单页面进入,参见图 3-19 和图 3-20。

3. 模板

这里的模板为查看所有订单页面和订单确认页面,没有单独的模板页面。

4. 接口测试

1) 测试用例

表 3-17 为删除订单的测试用例。设计两个测试用例:一个是删除总订单;另一个是删除单个订单。

表 3-17 删除订单的测试用例

编号	描述	期望结果
1	删除当前建立的单个订单	删除成功,且在显示页面中不显示
2	删除当前建立的总订单	删除成功,且在显示页面中不显示

2) XML 数据文件

"显示当前用户的所有订单"测试用例的测试数据在 orderConfig.xml 中定义如下。

```xml
...
<!--删除当前建立的单个订单 -->
<case>
    <TestId>order-testcase004</TestId>
    <Title>订单信息</Title>
    <Method>get</Method>
    <Desc>删除当前建立的单个订单</Desc>
    <Url>http://127.0.0.1:8000/delete_orders/0/1/</Url>
    <InptArg></InptArg>
    <Result>200</Result>
    <CheckWord>NOT,订单 id:0</CheckWord><!--检查单个订单是否被删除 -->
</case>
```

```
            <!--删除当前建立的总订单 -->
            <case>
                <TestId>order-testcase005</TestId>
                <Title>订单信息</Title>
                <Method>get</Method>
                <Desc>删除当前建立的总订单</Desc>
                <Url>http://127.0.0.1:8000/delete_orders/0/2/</Url>
                <InptArg></InptArg>
                <Result>200</Result>
                <CheckWord>NOT,上海市外滩一号</CheckWord><!--检查总订单是否被删除 -->
            </case>
</node>
```

3）测试代码

测试代码保持不变。

3.8 电子支付模块

电子支付模块包括使用支付宝、微信或其他手段进行支付，网上的资料已经很齐全了，读者也可以参阅参考文献[6]（注意，建立自己的电子支付，需要到网上申请支付宝或微信企业支付账号）。

3.9 建立自定义的错误页面

下面主要介绍如何建立自定义的 403、404、500 错误页面。首先建立 403.html、404.html 和 500.html，分别如图 3-21、图 3-22 和图 3-23 所示。

图 3-21　403.html

403.html 代码如下。

```
{%load staticfiles%}
<head>
    <meta charset="UTF-8">
    <title>403.html</title>
```

```
        <style type="text/css">
            * {
                margin: 0;
                padding: 0;
                background-color: #FFFFFF;
            }
            div{
                width: 700px;
                height: 200px;
                background-color: #FFFFFF;
            }
            .center-in-center{
                position: absolute;
                top: 40%;
                left: 30%;
            }
        </style>
</head>
<body>
    <div class="center-in-center">
        <img src="{% static 'image/403.JPG'%}" width="228" height="196" align="left">你没有这个权限!<br>
        <a href="/login_action/"><img src="{%static 'image/home.JPG'%}"></a>
    </div>
</body>
</html>
```

图 3-22　404.html

404.html 代码如下。

```
{%load staticfiles%}
<head>
    <meta charset="UTF-8">
    <title>404.html</title>
    <style type="text/css">
        * {
            margin: 0;
```

```
            padding: 0;
            background-color: #FFFFFF;
        }
        div{
            width: 700px;
            height: 200px;
            background-color: #FFFFFF;
        }
        .center-in-center{
            position: absolute;
            top: 40%;
            left: 30%;
        }
    </style>
</head>
<body>
<div class="center-in-center">
            <img src="{%static 'image/404.JPG'%}" width="228" height="196" align=
"left">你的页面找不到了<br>
            不如搜索一下你想要的或者刷新一下网页吧!<br>
            <img src="{%static 'image/reflesh.JPG'%}" onclick="location.reload(); ">
    </div>
</body>
</html>
```

图 3-23　500.html

500.html 代码如下。

```
{%load staticfiles%}
<head>
    <meta charset="UTF-8">
    <title>500.html</title>
    <style type="text/css">
        * {
            margin: 0;
```

```html
        padding: 0;
        background-color: #FFFFFF;
    }
    div{
        width: 700px;
        height: 200px;
        background-color: #FFFFFF;
    }
    .center-in-center{
        position: absolute;
        top: 40%;
        left: 30%;
    }
    </style>
</head>
<body>
    <div class="center-in-center">
        <img src="{% static 'image/500.JPG'%}" width="228" height="196" align="left">服务器内部错误,不能执行这个请求!<br>
        <a href="/login_action/"><img src="{% static 'image/home.JPG'%}"></a>
</div>
</body>
</html>
```

然后打开 settings.py,配置 templates 文件路径、关闭 Debug、配置 allowed_hosts。

```
...
DEBUG = False    #关闭 Debug
...
TEMPLATES = [
    {
        ...
        'DIRS': ['templates/'],   #配置 templates 文件路径
        ...
    },
]
...
ALLOWED_HOSTS=" * "
...
```

最后在 views.py 中做如下设置。

```
...
from django.shortcuts import render
def page_not_found(request):
```

```
        return render(request, '404.html')
def page_error(request):
    return render(request, '500.html')
def permission_denied(request):
    return render(request, '403.html')
...
```

这样,在页面中显示 403、404 和 500 号错误时就会显示对应的自定义网页。

第4章 构建安全的网站

4.1 密码的加密

2.3.2 节中提醒过大家,前面的代码是明文存储密码的,其实这是很危险的,这里将对密码进行 MD5 加密,以保证信息安全。在 goods/util.py 中定义加密方法 md5()如下。

```
...
import hashlib
...
#MD5 加密
    def md5(self,mystr):
        if isinstance(mystr,str):
            m = hashlib.md5()
            m.update(mystr.encode('utf8'))
            return m.hexdigest()
        else:
            return ""
...
```

注意:加密字符串 mystr 必须转为 bytes,才可以被加密。然后在注册和登录代码中分别调用该方法。

```
...
#用户注册
def register(request):
    ...
        #获取密码信息
        #加密 password
        password = util.md5(password)
...
#用户登录
def login_action(request):
    ...
        username = (request.POST.get('username')).strip()
```

```
            password = (request.POST.get('password')).strip()
            #加密 password
            password = util.md5(password)
...
#修改用户密码
def change_password(request):
...
        if request.method == "POST":
            #获取旧密码
            oldpassword=util.md5((request.POST.get("oldpassword", "")).strip())
            #获取新密码
            newpassword=util.md5((request.POST.get("newpassword", "")).strip())
            #获取新密码的确认密码
            checkpassword= util.md5((request.POST.get("checkpassword", "")).strip())
...
```

由于使用 MD5 对密码进行了加密,所以同样也需要对测试程序 interface/util.py 进行如下调整。

```
...
#初始化信息
    def inivalue(self,dataBase,ordertable,sign):
        ...
        #建立记录
        if(sign!="0"):         #sign=0,密码需要加密,否则不需要加密
            self.insertTable(dataBase,ordertable,values)
        #处理在用户注册的时候,需要将密码进行 MD5 加密处理
        else:
            dom =minidom.parse("initInfo.xml")
            self.root =dom.documentElement
            password =self.root.getElementsByTagName('password')
            password =str(password[0].firstChild.data).strip()
            md5password =self.md5(password)
            newvalues =values.replace(password,md5password)
            self.insertTable(dataBase,ordertable,newvalues)
        return values
...
```

4.2 防止 CSRF 攻击

4.2.1 CSRF 攻击介绍

跨站请求伪造(Cross-Site Request Forgery:CSRF)也被称为 One Click Attack 或者

session Riding,通常缩写为 CSRF 或者 XSRF,是一种对网站的恶意利用。听起来有点像跨站脚本(即 XSS,将在 4.4 中介绍),但它与 XSS 是不同的,XSS 利用的是站点内信任用户,而 CSRF 是通过伪装来自受信任用户的请求利用受信任网站。与 XSS 攻击相比,CSRF 不是流行的攻击方式,对其进行防范的资源也相当稀少,且难以防范,所以业界认为其比 XSS 更具危险性。

用一个 POST 请求做个比方,黑客可以构建自己的网页 form 界面,form 的 action 指向要攻击的网站,form 中元素的 name 与攻击网站的值保持一致,从而达到 CSRF 攻击的目的。

例如,被攻击的网站是 http://www.a.com,页面提交网站是 http://www.a.com/input.html,提交后处理的网站是 http://www.a.com/display.jsp。input.html 的网页内容如下。

```
...
<form action="display.jsp" method="post" >
地址:<input type="text" name="address" id="id_address" size="20" maxlength="100" required />
电话:<input type="text" name="phone" id="id_ phone" size="20" maxlength="100" required />
</form>
...
```

现在在本地构造一个界面冒充 input.html。

```
...
<form action="http://www.a.com/display.jsp" method="post" >
地址:<input type="text" name="address" id="id_address" size="20" required />
电话:<input type="text" name="phone" id="id_ phone" size="20"required />
</form>
...
```

这样,黑客就可以用自己的页面向 http://www.a.com/display.jsp 发起攻击了。在作者著作的《软件测试技术实战 设计、工具及管理》书的序言中曾经提及这么一件事情:

"2000 年,我所在的公司与 CCTV'开心辞典'节目组合作开发网上答题的项目,这是一个智力娱乐性节目,我编写了前端的答题代码,考虑到可能有人用计算机程序答题,如编写一个死循环,一直选择 B(或 A,或 C,或 D),这可以使答题的速度很快,命中率也非常高,为此我选用 JavaScript 过滤了使用死循环的答题者。可是,到了'开心辞典'正式使用这个软件的时候,发现仍然有人使用死循环答题,可我的程序是正确的。后来在一个聊天模块中通过登录账号找到了这位'达人',他说我们前端的确没有漏洞,他是通过自己编写的程序绕过我们前端进入系统后端的,而我们的后端并没有进行校验。当初如果有专业的测试人员,这个 Bug 是有可能避免的。"其实这就是一个很典型的 CSRF 攻击。

4.2.2　Django 是如何防范 CSRF 攻击的

在 2.3.2 节就介绍过 Django 是如何防范 CSRF 攻击机制的,而且 Django 默认是启动

CSRF 攻击机制的，本书前几章介绍的重点不在这里，所以关闭了 setting.py 中的这个开关。现在进入 setting.py 打开这个插件的开关。

```
...
MIDDLEWARE = [
    'django.middleware.security.SecurityMiddleware',
    'django.contrib.sessions.middleware.SessionMiddleware',
    'django.middleware.common.CommonMiddleware',
    'django.middleware.csrf.CsrfViewMiddleware',
    'django.contrib.auth.middleware.AuthenticationMiddleware',
    'django.contrib.messages.middleware.MessageMiddleware',
    'django.middleware.clickjacking.XFrameOptionsMiddleware',
]...
```

这样在所有模板中有表单提交(<form>…</form>)的地方都加上一个{% csrf_token %}标记。最后把 views.py 中的所有 render_to_response()方法用 render()方法替换(csrf 不支持 render_to_response()方法，正如 2.9.2 节中所述 render_to_response()逐步被 render()取代)。现在以登录模块分析 Django 是如何防范 CSRF 攻击的。在此之前，打开一个 HTTP 抓包工具，这里用的是 Fiddle 4，然后进入登录界面，查看网页源代码会发现如下代码。

```
...
<input type='hidden' name='csrfmiddlewaretoken' value=
'XltpK31i171tGLIH2leLWio0xM5TY8NC56oaU58CiIc5xLfqSiiehfJDSEnZesrX ' />
...
```

也就是说，{%csrf_token%}被一个名为 csrfmiddlewaretoken 的 hidden 类型取代了。其值为 XltpK31i171tGLIH2leLWio0xM5TY8NC56oaU58CiIc5xLfqSiiehfJDSEnZesrX 一个 100 位的字符串，然后查看 Fiddle 4，会看到页面产生了一个名为 csrftoken 的 cookie，其值也为 XltpK31i171tGLIH2leLWio0xM5TY8NC56oaU58CiIc5xLfqSiiehfJDSEnZesrX，如图 4-1 所示。

如果刷新这个登录页面，会发现这个字符串会发生相应的变化，但是 cookie 的值与 hidden 中的值永远保持一致。后来作者查询了一些资料，发现不仅 Django 使用这种方式处理 CSRF 注入，其他大部分系统都使用这种方法处理 CSRF 注入。在用户登录这个网站的时候产生一个叫作 csrf token(csrf 令牌)的随机字符串，即前面提到的 100 位会发生随机变化的字符串，然后把 csrf token 放入 cookie 中(所以，要是用 CSRF 防御机制，必须打开浏览器的 cookie)，并且放到页面的 form 表单中，产生一个类似<input type='hidden' name='csrfmiddlewaretoken' value='csrf token'的表单，最后在提交表单的时候验证 cookie 中的值是否与 hidden 的值保持一致，如果保持一致，则返回 200 代码，否则返回 403 拒绝访问代码，如图 4-2 所示。

4.2.3 针对 CSRF 防御接口测试代码的调整

为了适应增加对 CSRF 的防御功能，必须对测试代码进行调整，由于前面对代码进行

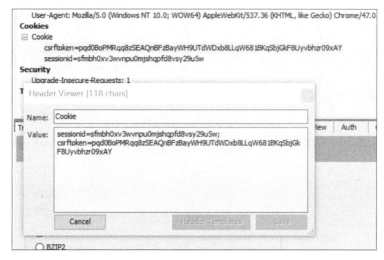

图 4-1　产生的 cookie

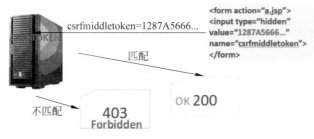

图 4-2　CSRF 防范示意图

了很好的封装,所以这里只调整 interface/util.py 中 Util 类中的 run_test() 方法就可以了。下面是改动后的代码。

```
...
#运行测试接口
    #mylist 测试数据
    #values 登录数据
    def run_test(self,mylist,values,sign):
        #获取测试 URL
        Login_url = self.url+"/login_action/"      #login_Url 为登录的 URL
        run_url = mylist["Url"]                     #run_url 为运行测试用例所需的 URL
        #获取 csrf_token
        data = self.s.get(Login_url)
        csrf_token = data.cookies["csrftoken"]
        #初始化登录变量
        #获取登录数据
        username = values.split(',')[1].strip("\"")
        password = values.split(',')[2].strip("\"")
        #判断当前测试是否需要登录
```

```
            if sign:
                #使用当前用户登录系统
                payload={"username":username,"password":password,"csrfmiddlewaretoken":
csrf_token}
                try:
                    data=self.s.post(Login_url,data=payload)
                except Exception as e:
                    print(e)
                #运行测试接口
                try:
                    #为POST请求,由于POST请求参数肯定是存在的,所以这里不判断有无参数
                    if mylist["Method"]=="post":
                        #有请求参数
                        payload=eval(mylist["InptArg"])
                        payload["csrfmiddlewaretoken"]=csrf_token
                        data=self.s.post(run_url,data=payload)
                    #为GET请求
                    elif mylist["Method"]=="get":
                        if mylist["InptArg"].strip()=="":
                            #没有请求参数
                            data=self.s.get(run_url)
                        else:
                            #有请求参数
                            payload=eval(mylist["InptArg"])
                            data=self.s.get(run_url,params=payload)
                except Exception as e:
                    print(e)
                else:
                    return data
...
```

(1) 通过代码 data = self.s.get(Login_url)访问登录页面。

(2) 通过代码 csrf_token = data.cookies["csrftoken"]获取产生的 CSRF 令牌 cookie。

(3) 在初始化登录操作与执行 POST 操作的时候把令牌参数 csrf_token 加入 POST 参数中。

在初始化登录操作中,代码如下。

```
...
#使用当前用户登录系统
payload={"username":username,"password":password,"csrfmiddlewaretoken":csrf_
token}
try:
    data=self.s.post(Login_url,data=payload)
...
```

执行 post 操作的代码如下。

```
...
# 为 POST 请求,由于 POST 请求参数肯定是存在的,所以这里不判断有无参数
if mylist["Method"] =="post":
    # 有请求参数
    payload =eval(mylist["InptArg"])
    payload["csrfmiddlewaretoken"] =csrf_token
    data =self.s.post(run_url,data=payload)
...
```

在 userInfoConfig.xml 中增加一个测试用例,测试不加载 csrftoken,程序会不会产生 403 返回码,并且返回的 text 中是否含有"Forbidden"字符串。

```
...
    <!--测试 CSRF 不启用,返回 403 码 -->
    <case>
        <TestId>userInfo-testcase006</TestId>
        <Title>修改用户密码</Title>
        <Method>post</Method>
        <Desc>密码修改成功</Desc>
        <Url>http://127.0.0.1:8000/change_password/</Url>
        <InptArg>{"oldpassword":"000000","newpassword":"123456","checkpassword":"123456"}</InptArg><!--新密码与旧密码不同,确认密码与新密码匹配 -->
        <Result>403</Result>
        <CheckWord>Forbidden</CheckWord>
    </case>...
```

在 interface/util.py 中的 run_test() 还要进行小小的改动。

```
...
    def run_test(self,mylist,values,sign):
        ...
        # 运行测试接口
        try:
            # 为 POST 请求,由于 POST 请求参数肯定是存在的,所以这里不判断有无参数
            if mylist["Method"] =="post":
                payload =eval(mylist["InptArg"])
                # 如果不是测试 CSRF 的
                if mylist["Result"]!="403":
                    payload["csrfmiddlewaretoken"]=csrf_token
                data =self.s.post(run_url,data=payload)
...
```

这里,如果返回码是 403,在请求参数中就不加入 csrfmiddlewaretoken 项。

4.3 权限操作的漏洞

试想,如果一个名为 Linda 的用户登录系统后可以通过 http://127.0.0.1:8000/update_address/1306/2/修改他的收货地址信息,另一个名为 Jerry 的用户登录系统后在浏览器地址栏中直接输入 http://127.0.0.1:8000/update_address/1306/2/也可以修改这条记录。这就产生了一个安全缺陷,解决这个缺陷的方法是在修改前先判断收货地址信息是否属于这个登录用户,如果不属于,就抛出错误提示信息,不进行相应的操作。在 goods/util.py 中加如下代码。

```
...
#通过 addressId 判断这个地址是否属于当前登录用户
    def check_User_By_Address(self,request,username,addressId):
        #获取 addressId 对应的 address 信息
        address = get_object_or_404(Address,id=addressId)
        #通过 username 获取对应的 user 信息
        user = get_object_or_404(User,username=username)
        #判断 address 对应的 user.id 与 username 获取的对应的 user.id 是否相等
        if address.user_id == user.id:
            return 1
        else:
            return 0
...
```

然后修改 view.py 中的方法 update_address()。

```
...
def update_address(request,address_id,sign):
    util = Util()
    username = util.check_user(request)
    if username == "":
        uf = LoginForm()
        return render(request,"index.html",{'uf':uf,"error":"请登录后再进入"})
    else:
        #判断修改的地址是否属于当前登录用户
        if not util.check_User_By_Address(request,username,address_id):
            return render(request,"error.html",{"error":"你试图修改不属于你的地址信息!"})
        else:
            #获取指定地址信息
            address_list = get_object_or_404(Address, id=address_id)
...
```

粗体字部分用于判断修改的地址是否属于当前登录用户。这里建立一个名为 error.

html 的模板文件。

```html
{%extends "base.html" %}
{%block content %}
                    <ul class="nav navbar-nav navbar-right">
                        <li><a href="/user_info/">{{user}}</a></li>
                        <li><a href="/logout/">退出</a></li>
                    </ul>
                </div><!--/.nav-collapse -->
            </div>
        </nav>
        <div class="row" style="margin-top:30px">
            <div class="col-lg-6">
                <div class="input-group">
                </div><!--/input-group -->
            </div><!--/.col-lg-6 -->
        </div><!--/.row -->

        <div class="container theme-showcase" role="main">
            <font color="#FF0000">{{error}}</font>
        </div><!--/container glyphicon glyphicon-phone border-style:none; -->
{%endblock %}
```

出错信息提示如图 4-3 所示。

图 4-3 出错信息提示

对于收货地址的删除操作，也加如下代码。

```
...
def delete_address(request,address_id,sign):
    util =Util()
    username =util.check_user(request)
    if username=="":
        uf =LoginForm()
        return render(request,"index.html",{'uf':uf,"error":"请登录后再进入"})
    else:
        if not util.check_User_By_Address(request,username,address_id):
            return render(request,"error.html",{"error":"你试图删除不属于你的地址信息!"})
        else:
...
```

这里可以在 addressConfig.xml 中加两条测试数据,分别测试"试图修改不属于自己的地址信息"和"试图删除不属于自己的地址信息"。

```xml
...
    <!--试图修改一个不属于自己的地址 -->
    <case>
        <TestId>address-testcase006</TestId>
        <Title>地址信息</Title>
        <Method>post</Method>
        <Desc>试图修改一个不属于自己的地址</Desc>
        <Url>http://127.0.0.1:8000/update_address/100/1/</Url><!-- "100"作为测试程序地址 id 插入数据库表中 -->
        <InptArg>{"address":"上海市延安中路100号","phone":"13681166561"}</InptArg>
        <Result>200</Result>
        <CheckWord>你试图</CheckWord><!--检查"试图修改不属于自己的地址信息"有无得逞 -->
    </case>
    <!--试图删除不属于自己的地址信息 -->
    <case>
        <TestId>address-testcase007</TestId>
        <Title>地址信息</Title>
        <Method>get</Method>
        <Desc>删除地址信息</Desc>
        <Url>http://127.0.0.1:8000/delete_address/100/1/</Url><!-- "100"作为测试程序地址 id 插入数据库表中 -->
        <InptArg></InptArg>
        <Result>200</Result>
        <CheckWord>你试图</CheckWord><!--检查"试图删除不属于自己的地址信息"有无得逞 -->
    </case>
...
```

然后再改动测试程序 addressTest.py。

```python
...
        #初始化非法操作信息
        self.myuservalue = ""
        self.addressvalues = ""

    #开始测试
    def test_address_info(self):
...
        #对非法操作进行的测试
        if ("你试图" in mylist["CheckWord"]):
            #建立用户信息
```

```
        self.myuservalue = "1,\"121\",\"123456\",\"12345@126.com\""
        #从mylist["Url"]中获取关联地址信息对应的用户id
        id = (mylist["Url"]).split("/")[4]
        address="淮海中路"
        self.addressvalues=id+",'"+address+"','13666666666',1"
        #建立一个用户
        self.util.insertTable(self.dataBase,self.userTable,self.myuservalue)
        #建立一个地址
        self.util.insertTable(self.dataBase,self.addressTable,
        self.addressvalues)
...
    def tearDown(self):
        #对非法操作的处理
        self.util.tearDown(self.dataBase,self.userTable,self.myuservalue)
        self.util.tearDown(self.dataBase,self.addressTable,self.addressvalues)
        #清除其他初始化信息
        ...
...
```

在删除单个订单、删除总订单里面也会出现这样的问题。首先，在 goods/util.py 中加如下代码。

```
...
#通过 orderId 判断这个订单是否属于当前登录用户
    def check_User_By_Order(self,request,username,orderId):
        #获取 orderId 对应的 order 信息
        order = get_object_or_404(Order,id=orderId)
        #通过 username 获取对应的 user 信息
        user = get_object_or_404(User,username=username)
        #判断 address 对应的 user.id 与 username 获取的对应的 user.id 是否相等
        if order.user_id == user.id:
            return 1
        else:
            return 0

#通过 ordersId 判断这个地址是否属于当前登录用户
    def check_User_By_Orders(self,request,username,orderId):
        #获取 orderId 对应的 orders 信息
        orders = get_object_or_404(Orders,id=orderId)
        #获取 orders.id 对应的 order 信息
        order = Order.objects.filter(order_id=orders.id)
        #通过 username 获取对应的 user 信息
        user = get_object_or_404(User,username=username)
        #判断 address 对应的 user.id 与 username 获取的 user.id 是否相等
```

```
            if len(order)>0:
                if order[0].user_id ==user.id:
                    return 1
                else:
                    return 0
...
```

然后改动产品代码 views.py 中的 delete_orders()方法。

```
...
def delete_orders(request,orders_id,sign):
        #如果删除单独一个订单
        if sign =="1" or sign=="3":
            #判断修改的地址是否属于当前登录用户
            if not util.check_User_By_Order(request,username,orders_id):
                return render(request,"error.html",{"error":"你试图删除不属于你的单独一个订单信息!"})
            else:
....
        elif sign =="2":
            if not util.check_User_By_Orders(request,username,orders_id):
                return render(request,"error.html",{"error":"你试图删除不属于你的总订单信息!"})
            else:
...
```

最后设计测试数据与测试程序。在 orderConfig.xml 中数据如下。

```
...
    <!--试图删除当前不属于自己的总订单 -->
    <case>
        <TestId>order-testcase006</TestId>
        <Title>订单信息</Title>
        <Method>get</Method>
        <Desc>试图删除当前不属于自己的总订单</Desc>
        <Url>http://127.0.0.1:8000/delete_orders/100/2/</Url><!--"100"作为测试程序总订单 id 插入数据库表中 -->
        <InptArg></InptArg>
        <Result>200</Result>
        <CheckWord>你试图</CheckWord><!--检查"试图删除不属于自己的总订单"有无得逞 -->
    </case>
    <!--试图删除当前不属于自己的单个订单 -->
    <case>
        <TestId>order-testcase007</TestId>
```

```xml
            <Title>订单信息</Title>
            <Method>get</Method>
            <Desc>试图删除当前不属于自己的单个订单</Desc>
            <Url>http://127.0.0.1:8000/delete_orders/100/1/</Url><!--"100"作为测试
程序单个订单 id 插入数据库表中 -->
            <InptArg></InptArg>
            <Result>200</Result>
            <CheckWord>你试图</CheckWord><!--检查"试图删除不属于自己的单个订单"有无
得逞 -->
        </case>
...
```

在 orderTest.py 中,代码如下。

```
...
#初始化非法操作信息
        self.myuservalue = ""
        self.myordervalues = ""
        self.myordersvalues = ""

        #开始测试
    def test_order_info(self):
        for mylist in self.mylists:
            if ("你试图" in mylist["CheckWord"]):
                #建立用户信息
                self.myuservalue = "1,\"121\",\"123456\",\"12345@126.com\""
                #建立用户信息与上面和用户关联的单个订单和总订单,id 从 mylist["Url"]
中获取
                id = (mylist["Url"]).split("/")[4]
                self.myordervalues = id+",2,0,"+id+",1"
                self.myordersvalues = id+",\"Sept. 13, 2017, 3:55 a.m.,\",0,0"
                #建立一个用户
                self.util.insertTable (self. dataBase, self. userTable, self.
myuservalue)
                #建立一个总订单
                self.util.insertTable (self. dataBase, self. ordersTable, self.
myordersvalues)
                #建立一个单个订单
                self.util.insertTable (self. dataBase, self. orderTable, self.
myordervalues)
                data = self.util.run_test(mylist,self.userValues,self.sign)
...
    def tearDown(self):
            #对非法操作的处理
```

```
            self.util.tearDown(self.dataBase,self.userTable,self.myuservalue)
            self.util.tearDown(self.dataBase,self.orderTable,self.myordervalues)
            self.util.tearDown(self.dataBase,self.ordersTable,self.myordersvalues)
    ...
```

4.4 防止 XSS 攻击

在百度百科中，XSS 攻击是这样定义的："XSS 攻击的全称是跨站脚本攻击，是为了不和层叠样式表(Cascading Style Sheets,CSS)的缩写混淆，故将跨站脚本攻击缩写为 XSS。XSS 是一种在 Web 应用中的计算机安全漏洞，它允许恶意 Web 用户将代码植入提供给其他用户使用的页面中。" XSS 攻击注入包括持久型、反射型和 DOM 型。最典型的一个例子是在文本框中输入一段 JavaScript 语句，然后在页面显示的时候这个 JavaScript 语句被激活执行。最简单的一个例子是，在收货地址输入栏中输入＜img src＝"javascript:alert('hi')"＞，显示的时候看看是否 JavaScript 被执行。这个测试用 XML 实现比较困难，因为 XML 中不允许存在 HTML 中的特殊字符，如＜、＞、"，然而，用 <、>或 quot;替代意义就不大了。经过手工测试，发现结果非常令人满意，Django 框架已经帮助实现了对 XSS 注入的防范。

4.5 防止 SQL 注入

在百度百科中，SQL 注入是这样定义的："所谓 SQL 注入，就是通过把 SQL 命令插入 Web 表单提交或输入域名或页面请求的查询字符串，最终达到欺骗服务器执行恶意的 SQL 命令。具体来说，它是利用现有应用程序，将(恶意)SQL 命令注入后台数据库引擎执行的能力，它可以通过在 Web 表单中输入(恶意)SQL 语句得到一个存在安全漏洞的网站上的数据库，而不是按照设计者的意图执行 SQL 语句。例如，先前的很多影视网站泄露 VIP 会员密码大多就是通过 Web 表单递交查询字符暴出的，这类表单特别容易受 SQL 注入式攻击。"

其实，3.4.3 节中有一个测试用例就是用来测试是否存在 SQL 注入，在模糊查询时，SQL 语句往往是这样的：select * from table where title like '％var％'，其中 var 是用户输入的字符，在 goods-testcase005 中输入的 var 是"％"，如果程序没有进行任何处理，这个 SQL 语句就变成了 select * from table where title like '％％％'。这样，table 表中的所有记录就都被查询出来了。在程序中没有进行任何处理，说明 Django 框架自动处理了这个注入。

除了"％"的注入，在用户登录时候的 SQL 注入更加危险，正如产品代码中，判断用户是否合法，类似的 SQL 语句是这样的：select * from goods_user where username= 'usernamevar' and password='passwordvar'，其中 usernamevar 与 passwordvar 是通过前端输入的，如果返回的结果不为空，则认为用户合法，否则认为不合法。设想，如果 usernamevar＝ 111，passwordvar＝ ' or 1＝1 --',SQL 语句就变为 select * from goods_user where username＝

'111' and password='' or 1=1 --'',因为 1=1 是永远正确的,又由于前面是 or 操作,所以这条 SQL 语句的返回记录是不为空的。因此,在 loginRegConfig.xml 中设计如下的测试数据。

```xml
...
    <!--SQL 注入测试 -->
    <case>
        <TestId>loginReg-testcase007</TestId>
        <Title>用户登录</Title>
        <Method>post</Method>
        <Desc>SQL 注入测试</Desc>
        <Url>http://127.0.0.1:8000/login_action/</Url>
        <InptArg>{"username":"111","password":"' or 1=1 -- '"}</InptArg><!--用户名、密码与 initInfo.xml 中的用户信息相同 -->
        <Result>200</Result>
        <CheckWord>用户名或者密码错误</CheckWord>
    </case>
...
```

运行测试程序 loginRegTest.py,测试通过,说明 Django 也已经处理了这种情况的 SQL 注入。

参 考 文 献

[1] 虫师. Web 接口开发与自动化测试 基于 Python 语言[M]. 北京：电子工业出版社,2016.
[2] 虫师. Selenium 2 自动化测试实战 基于 Python 语言[M]. 北京：电子工业出版社,2017.
[3] 齐伟. 跟着老齐学 Python 从入门到精通[M]. 北京：电子工业出版社,2016.
[4] 齐伟. 跟着老齐学 Python Django 实战[M]. 北京：电子工业出版社,2017.
[5] 何敏煌. Python 新手使用 Django 建站的 16 堂课[M]. 北京：清华大学出版社,2017.
[6] hornbills. 支付宝即时到账接口的 python 实现,示例采用 django 框架[EB/OL].[2014-10-21]. http://blog.csdn.net/hornbills/article/details/40338949.
[7] Java 虾米的博客. 细说 Python 2.x 与 3.x 版本区别[EB/OL].[2017-05-02]. http://www.cnblogs.com/wangyayun/archive/2017/05/02/6794611.html.